"*Accountable* is striking in its clarity and insights born from the authors' experience as investors on the front lines of capitalism's excesses and potential. The case studies of some of the best-known companies and investors don't pull punches while offering tangible examples of the changes in culture and rules of the game required for a new type of capitalism."

—*Daniella Ballou-Aares, CEO and cofounder of the Leadership Now Project*

"If we don't contend with climate change, inequality, and the generational challenges tearing at our democracy, we will destroy our economy. *Accountable* offers a different vision. Unwilling to settle for easy answers or superficial changes, O'Leary and Valdmanis push us all to ask more of our economic system."

—*Senator Michael F. Bennet*

"If we want to save free-market enterprise economics and all the benefits it brings, we have to reform capitalism and the way corporations behave. This was a big part of the modern compassionate Conservative agenda I championed—and what this book is all about. The authors do a great job in explaining that this is not a wealth-bashing, negative agenda but a positive and exciting one. Businesses linking better with their communities. Impact investing to solve social problems. Companies stepping up and recognizing their wider responsibilities beyond simply maximizing profits. Business doing good is good business—and this book puts that beyond doubt."

—*David Cameron, former prime minister of the United Kingdom*

"Solving our biggest social and environmental problems means reshaping the role that businesses play in our society. This informative book addresses the challenges we face in achieving this crucial transformation."

—*Sir Ronald Cohen, chair of the Global Steering Group for Impact Investment (GSG)*

"In today's reality, business represents a picture of dysfunctionality, of excesses benefitting the few, of gross injustice. Tomorrow's business leaders have an obligation to create a different reality reflecting our better hopes and inspirations. *Accountable* takes you on that journey. It's a wonderful trip."

—*Peter Georgescu, chairman emeritus of Young & Rubicam*

"For decades, the dominant view has been that the purpose of companies is to make money for their shareholders. No more. Combining compelling examples and analytical insights, O'Leary and Valdmanis explain why the world is changing, and provide a road map for how shareholders and citizens can and must transform the corporate landscape to save capitalism from itself."

—*Oliver Hart, Harvard University, 2016 Nobel laureate in economics*

"*Accountable* takes a fresh look at a fundamental question of capitalism—whose interests should corporations serve. Its call for collective action by consumers, workers, and investors to reorient corporate behavior may foretell the next decade."

—*Jonathan Levin, dean of Stanford Graduate School of Business*

"*Accountable* reminds us that today's short-termism is as dangerous for corporations as it is for the world and uses a combination of great stories and thoughtful analysis to suggest that we must find a way to change the purpose of our corporations if we are to build a society that works for all of us. I enjoyed it enormously."

—Rebecca M. Henderson, John and Natty McArthur University Professor at Harvard University

"Even most capitalists know that capitalism, as practiced in the United States in recent decades, has a lot to answer for. O'Leary and Valdmanis have given us thoughtful, well-researched, and compelling ways to rethink how to assure that prosperity and fairness are linked."

—Deval Patrick, former governor of Massachusetts

"*Accountable* presents a fresh, balanced, highly readable, and deeply informed case for how the pursuit of sustained financial success and the exercise of social responsibility to employees, consumers, and society not only can—but must—go hand in hand if we are to have the world we seek in the future. Especially at a time when people's trust in all institutions, including business, are at long-term lows, we must renew the purpose of business itself to recognize that it has a responsibility—and opportunity—to use its resources—money, people, and energy—for the long-term benefit of society as well as the company. As *Accountable* shows us, this will build increased loyalty and retention of employees and customers, keys to sustainable financial progress."

—John Pepper, former chairman and CEO of P&G

"The issues posed could not be more topical or provocative. The coronavirus pandemic has catapulted capitalism into the future, where we need new and better answers to the crucial questions 'What matters most in our society?' 'How should we organize work to support those priorities?' 'What's a fair way to pay people for their contributions?' *Accountable* offers a fresh and compelling perspective on these questions."

—David Roux, founder of Silver Lake

"Thought-provoking and insightful, *Accountable* offers a pragmatic and original road map to transform capitalism into a system that's more inclusive, sustainable, and just. More than ever before, this is the book our economy needs."

—Dr. Rajiv J. Shah, president of the Rockefeller Foundation

"For many years, the authors have been at the forefront of thinking on how markets and business can better serve society. *Accountable* captures not only the history that underpins modern-day corporate purpose, but its present and future. The central message—that capitalism can and must be part of the solution to society's greatest challenges—has never been more important."

—Martin Whittaker, CEO of JUST Capital

ACCOUNTABLE

ACCOUNTABLE

THE RISE OF CITIZEN CAPITALISM

MICHAEL O'LEARY

AND WARREN VALDMANIS

An Imprint of HarperCollins*Publishers*

 For information, address HarperCollins Publishers, 195 Broadway, New York, NY 10007.

HarperCollins books may be purchased for educational, business, or sales promotional use. For information, please email the Special Markets Department at SPsales@harpercollins.com.

FIRST EDITION

Library of Congress Cataloging-in-Publication Data
Names: O'Leary, Michael, 1989– author. | Valdmanis, Warren, author.
Title: Accountable : the rise of citizen capitalism / Michael O'Leary and Warren Valdmanis.
Description: New York : Harper Business, 2020. | Includes bibliographical references and index. | Identifiers: LCCN 2020000593 (print) | LCCN 2020000594 (ebook) | ISBN 9780062976512 (hardcover) | ISBN 9780062976550 (ebook)
Subjects: LCSH: Capitalism. | Free enterprise. | Disinvestment. | Social responsibility of business. | Citizenship.
Classification: LCC HB501 .O494 2020 (print) | LCC HB501 (ebook) | DDC 174/.4—dc23
LC record available at https://lccn.loc.gov/2020000593
LC ebook record available at https://lccn.loc.gov/2020000594

20 21 22 23 24 LSC 10 9 8 7 6 5 4 3 2 1

In memory of our fathers

They went profoundly into the science of business, and indicated that the purpose of manufacturing a plow or a brick was so that it might be sold. To them, the Romantic Hero was no longer the knight, the wandering poet, the cowpuncher, the aviator, nor the brave young district attorney, but the great sales-manager . . . who devoted himself and all his young samurai to the cosmic purpose of Selling—not of selling anything in particular, for or to anybody in particular, but pure Selling.

—Sinclair Lewis, *Babbitt* (1922)

"I got to figure," the tenant said. "We all got to figure. There's some way to stop this. It's not like lightning or earthquakes. We've got a bad thing made by men, and by God that's something we can change."

—John Steinbeck, *The Grapes of Wrath* (1939)

CONTENTS

ACCOUNTABLE

INTRODUCTION

SAVING CAPITALISM FROM ITSELF

DIVISION, DESPAIR, AND DIVIDENDS IN CORPORATE AMERICA

If all else fails, Peter Thiel has an escape route.

The billionaire cofounder of PayPal has hedged himself against the collapse of society. His answer? New Zealand. Thiel purchased a five-hundred-acre sheep farm in South Island, a comfortable seven thousand miles away from his home in California. There he's protected from crisis, pandemic, and a population increasingly on the brink—of what? That's the worry.

He's not alone. Drawn to the country's six-to-one sheep-to-human ratio and neutral stance on just about everything, Thiel and his Silicon Valley cohort have contributed to a tenfold increase in applications for New Zealand's "Investor Plus" visas—which require a three-year investment of over $6 million—from 2010 to 2018. LinkedIn's cofounder Reid Hoffman estimates that half of his billionaire friends have apocalypse insurance. "Saying you're 'buying a house in New Zealand' is kind of a wink, wink, say no more."

For those hoping to stay closer to home, $4.5 million will buy a "Survival Condo Penthouse unit" in a converted missile silo in Kansas. The unit boasts nine-foot-thick walls and military-grade security. Residents can schedule airport pickup on the Pit-Bull VX armored truck. At the silo, they can stroll in the dog park, work up a sweat on the climbing wall, or watch the end of times from the comfort of their own movie theater. One customer wrote, "I feel better knowing that I have a luxury survival bunker for my family if anything happens."

For today's wealthy, the new must-have is a way out.

It feels somehow appropriate that a corporation would aim to profit from the end of capitalism. Alexis de Tocqueville noted that Americans always "put something heroic into their way of trading." He wrote, "I know of no country, indeed, where wealth has taken a stronger hold on the affections of men."

Unique in world history, America is a nation founded by corporations—two corporations, in fact: the Massachusetts Bay Company and the Virginia Company, both for-profit ventures before becoming political colonies. Half the passengers on the Mayflower were not Pilgrim church members but were conscripted by investors.

Before thirteen colonies became one nation, seven hundred English shareholders became one company. A single share in the Virginia Company cost 12 pounds, 10 shillings—six months' wages for an ordinary worker. For those willing to work seven years for the company in Virginia, the share was free.

As the historian Bhu Srinivasan writes, those early American settlements were funded through private investment, "with the risks borne by private interests and the lives of men."

The lives of men indeed: After seventeen years and hundreds of supply ships, 85 percent of Virginia's settlers had died. So King James I revoked the company's charter in 1624, forcing Virginia to be ruled as a colony instead.

Those were inauspicious beginnings for a country that would win the Cold War 365 years later in the name of capitalism. And though those corporate beginnings have transmogrified into Thanksgiving school plays, the United States has always struggled with the morality of its economy. This is a country built off tobacco and slaves, fossil fuels and global finance. Corporations had legal personhood before women did.

Rutherford B. Hayes, president during the Gilded Age, wrote in his diary, "This is a government of the people, by the people, and for the people no longer. It is government by the corporations, of the corporations, and for the corporations." The country's history has always been an uncomfortable balance of private and public interests, of people and the corporations they create.

Capitalism is once again on trial. Pernicious inequality and catastrophic climate change. Global instability and plateauing growth. Decaying institutions and receding opportunity.

Inequality is at a postwar high. In Boston, the gap in life expectancy between those living in adjacent rich and poor neighborhoods is *thirty-three years.* Growth in urban and rural areas has diverged. Since 2010, the largest fifty-three metro areas have accounted for over 90 percent of the nation's population growth. Half of rural counties have lost population since 2000. Across the country, only one in three Americans believes their children will "be better off financially than their parents." Meanwhile, according to one survey of professionals, half of Americans don't feel a connection to their company's mission or get a sense of meaning or significance from their work.

We often talk of stagnant wages but not what those wages can buy. Whereas the manufactured bread and circuses of fast food and flat-screen televisions have gotten cheaper, the things that give our lives dignity—education, fresh food, health care—have gotten much more expensive. If we were to adjust wage growth by a Consumer Dignity Index instead of the standard Consumer Price Index, we'd see that most people's wages haven't been stagnant; they've been in steep decline.

We've descended into a corporate-sponsored Brave New World, complete with a calcified caste system and the *soma* of cheap consumer goods and drugs—licit or illicit—to dull the pain. This is how the world ends: not with a bang but with a binge-worthy Netflix series.

Based on one 2020 study, 56 percent of people worldwide believe that "capitalism as it exists today does more harm than good in the world." Only half of American adults under forty view capitalism favorably—down from two-thirds in 2010. Four in ten Americans say they would prefer living in a socialist country over a capitalist one. It's all enough to make us ask: Can we save capitalism from itself? Should we?

The United States was born by the corporation, and it may die by it, too. Maybe Thiel isn't crazy to want a way out.

THE BOOK IN SUMMARY

Our most critical social and environmental challenges have been caused, in large part, by the explicit amorality of corporations. Our economy is dominated by the ideology that private vice makes for public virtue. And so our world is dominated by corporations that reflect no deeper purpose than profit. They embody no broader values than shareholder value. They are accountable to nothing but the bottom line.

Transforming our corporations will be hard. And we can't fix all that is wrong with capitalism without changing government and individual behavior as well. But neither can we make real progress without addressing the lack of purpose, values, and accountability within our corporations.

Capitalists should want to fix corporations. Too often *profit* is confused with *value*, but they are not the same thing. Profit is what happens today. Value is the long-term prosperity driven by strong customer, employee, and supplier relationships; positive community involvement; sustainable production; and cooperation with government in problem solving. We will make our corporations more valuable when we focus them on these things.

Furthermore, capitalists should want to fix corporations because capitalists are citizens, too. Before the pandemic, corporate profits were higher than ever, but workers, society, and the planet still suffered. This can't go on forever, and it won't. Capitalists, in their capacity as citizens, have an interest not just in avoiding the guillotine but also in living in a more equal, healthy, and sustainable world. That requires asking fundamental questions about our corporations: What is their purpose? What are they capable of producing for society?

"Hegel predicted that the basic unit of modern society would be the state . . . Lenin and Hitler that it would be the political party," wrote John Micklethwait and Adrian Wooldridge in their book *The Company: A Short History of a Revolutionary Idea.* "Before that, a succession of saints and sages claimed the same for the parish church . . . they have all been proved wrong. The most important organization in the world is the company: the basis of the prosperity of the West and the best hope for the future of the rest of the world."

Many critics of capitalism focus almost entirely on political philosophy and public policy. Thomas Piketty's *Capital in the Twenty-first Century* has more index entries on the French Revolution than it does on corporations. We risk missing a fundamental fact: that capitalism today is made up of corporations. Big corporations.

Think of how you spent your morning. You may have woken up to a tune on your iPhone (Apple, the number 4 largest corporation in the United States), with cell service powered by AT&T (number 9). Your morning medicine may have come from CVS (number 7), distributed by McKesson (number 6), and paid for by UnitedHealthcare (number 5). You may have driven a GM (number

10) car to work—insured by Geico (owned by Berkshire Hathaway, number 3) and fueled at ExxonMobil (number 2). On the way, you may have passed a Walmart (number 1) and a Whole Foods (owned by Amazon, number 8). And that's it, the top ten—even before you check Google in line at Starbucks while taking out your Bank of America credit card.

As the nineteenth-century historian and statesman George Bancroft put it, "Commerce defies every wind, outrides every tempest, and invades every zone." What began as a formal way to organize people and capital around a common cause has become Wells Fargo and Facebook, the daily nine to five and the McDonald's Dollar Menu. The marketplace—capitalism incarnate—is not some place we go. It's the air we breathe.

The first three chapters of this book are the story of fiduciary absolutism—a distorted and extreme version of fiduciary duty that binds corporate leaders to chasing short-term profit.

Corporations are a great modern multiplier of human effort. As we'll see in chapter 1, they have power for both good and ill. They can heal the sick, feed the hungry, and house the homeless. But they are also enmeshed in all our worst social and environmental problems. Inequality, disease, pollution—all inextricably bound up with our largest corporations. If there is something broken about capitalism today, there must be something broken at our largest corporations.

Corporations are capable of achieving so much more than they do today. Many are failing their employees, customers, communities, and other stakeholders. They're even failing their shareholders—who are better served when corporations focus on creating long-term value. As we'll see in chapter 2, that requires commitment to improving the lives of your customers, building trust with your employees, sustaining your community, stewarding your environment, and working collaboratively with the government.

Profits at our public companies may have set records, but society still suffered. This is usually blamed on shareholders. In chapter 3, we'll see how *most* shareholders would benefit far more if all corporations were transformed into long-term, stakeholder-oriented companies driven by a deeper purpose.

The next four chapters show reformers struggling against the conflicting demands of earning a profit for shareholders and doing good for society.

Many of the current reforms risk becoming hollow victories. In chapter 4, we'll see Goldman Sachs' 10,000 Women initiative, Coca-Cola's partnership with the Special Olympics, and Hyundai's funding for cancer research. Nearly nine in ten corporations in the S&P 500 now issue sustainability reports, up from only 20 percent in 2011. These are all good efforts that promise progress. But are they enough?

The New York City Pension Fund has divested itself from oil and gas companies, joining more than a thousand other institutions representing $11 trillion in assets. That feels like a victory, but in chapter 5 we'll ask: Does it matter? Does divestment work?

Impact investing has become a $500 billion industry—an industry we'll examine in chapter 6. The United Nations announced that $80 trillion is now covered by its Principles for Responsible Investment (PRI). But at the same time, two-thirds of Americans don't believe the financial system benefits the economy today. Our thousand largest corporations are owned 73 percent by institutional investors—mostly mutual funds concerned more with quarterly earnings than with long-term prosperity. As one political scientist summarized, "Our economy is dominated by property unowned by natural individuals." Can our financial system be saved by socially responsible investors?

And all the while, CEOs' pay has increased by 940 percent since 1978 while S&P 500 companies spent the last decade buying back

$4 trillion of their stock—$4 trillion that could have been invested in the workforce. In chapter 7 we'll ask: Why can't government just force our corporations to be good?

The final two chapters detail a different vision, one we'll call citizen capitalism. Getting there will require one thing above all else: accountability. Accountability for the corporations we create; accountability to all the stakeholders they touch; accountability for the world we will leave our children and grandchildren; accountability, ultimately, to one another.

In chapter 8 we'll offer our own proposal: corporations built around a deeper social purpose, held accountable to that purpose, and rebuilt to align their purpose with their prosperity. In chapter 9, we'll see how buyers, workers, savers, and voters are already influencing corporations to better reflect their values.

Our book isn't a naive plea for corporations to be more responsible; it's an unsentimental blueprint for how to build an economy that generates prosperity without peril. The journey starts with corporations and, in the end, depends on all of us.

We're not academics or journalists critiquing capitalism from the sidelines. We've lived in its engine room, investing billions of dollars and working across six continents and a dozen industries. We earned our MBAs from schools where the virtues of capitalism are recited chapter and verse. But over the last half decade, we've worked at the forefront of reform, helping to launch a social impact investment fund that has invested hundreds of millions of dollars in purpose-driven companies focused on benefiting society. We can debunk the myths of capitalist reform from the inside.

We've come to the conclusion that we need to ask more of capitalism. No one minds when great corporations serve their consumers well and keep some of the gains for their shareholders—that's the American way. What people mind is the economics of extraction, the kind of capitalism in which corporations maximize their short-term profits at any cost.

THE ARGUMENT FOR THE DEFENSE

Some people are ready to tear the whole of capitalism down. The Icelandic rock band Hatari ("Haters") has been described variously as "IcePop," "pseudo-fascist cyborgs," and a "bondage techno performance art group." It announced in 2018 that it was disbanding because it had failed to meet its stated goal of ending capitalism but came out of retirement in 2019 to enter the popular Eurovision music competition. It used its international recognition to challenge Israeli prime minister Benjamin Netanyahu to *glima*—described as "Icelandic trouser-grip wrestling." If Hatari won, it would get "the first ever Hatari-sponsored liberal BDSM colony on the Mediterranean coast." If it lost, the Israeli government would get "full political and economic control of South-Icelandic Island municipality Vestmannaeyjar."

Hatari self-consciously represents the extremes of anticapitalism, but there are others with a more staid—though no less ambitious—approach. The activist and environmentalist Naomi Klein suggested that climate change provides a great opportunity to end capitalism:

> Our economic system and our planetary system are now at war. Or, more accurately, our economy is at war with many forms of life on earth, including human life. What the climate needs to avoid collapse is a contraction in humanity's use of resources; what our economic model demands to avoid collapse is unfettered expansion. Only one of these sets of rules can be changed, and it's not the laws of nature.

Only one in four people trusts corporate CEOs to right society's wrongs, and only one in five trusts them to be truth-tellers and overall ethical. Many businesspeople don't even trust themselves. In one survey of business school students, only 60 percent trust corporations to do good things for people in the long run.

It's easy to forget why capitalism is worth saving. When colonists arrived in Massachusetts, they initially agreed to grow their food collectively. Unfortunately, that method of production didn't work. It foreshadowed a common joke about socialism: "They pretend to pay us, and we pretend to work." And so the colonists privatized. As the governor of Massachusetts explained, "This had very good success, for it made all hands very industrious. . . . The women now went willingly into the field, and took their little ones with them to set corn; which before would allege weakness and inability."

Since then, capitalism's 250-year report card reads surprisingly well.

During the three millennia preceding 1750, economic growth per person averaged around 0.01 percent per year. Forget doing better than your parents; for most of human history, you couldn't do better than your ancestors. Since 1750, however, the world's economic output per person has skyrocketed by more than thirty-seven times. Just since 1990, we have lifted a billion people from poverty. That didn't happen by accident; it happened by capitalism.

"I weigh my words," said Nicholas Murray Butler, the president of Columbia University for most of the first half of the twentieth century, "when I say that in my judgment the limited liability corporation is the greatest single discovery of modern times, whether you judge it by its social, by its ethical, by its industrial or, in the long run . . . by its political, effects. Even steam and electricity are far less important than the limited liability corporation and would have been reduced to comparative impotence without it."

Capitalism has created spectacular economic growth, eradicating disease and giving billions of people a chance to live a life beyond meager subsistence in the process. For all its harms, we must never forget that our economic system has also created tremendous prosperity. In 2019, Darren Walker, the head of the Ford Foundation, took a balanced view in a letter entitled "In Defense of Nuance." "We can see how our capitalist systems have broken down, *while also* appreciating that markets have helped reduce the number of people around the globe who live in poverty," he wrote. ". . . Within this kind of rationality—within this kind of complexity—I believe we can find reason for hope."

A world without corporations, free markets, and private ownership wouldn't restore balance to our economy; it would destroy the tools of progress. Capitalism has changed over time, but it is always marked by private ownership of most assets, competition among companies, and prices adjusted to allocate scarce resources. Where prosperity exists in our world today, it can be traced back to this potent combination. Even the relatively socialist countries of northern Europe are still clearly capitalist by this standard.

Nonetheless, the test of capitalism is not its past track record but its ability to meet our new and pressing social and environmental challenges. Many corporations today are too obsessed with creating short-term profit for shareholders to deliver the changes society is demanding. Fortunately, we don't need to change every one of America's 6 million companies. But we do need to change the Fortune 500. According to Fortune.com, these companies represent two-thirds of US GDP, $1.1 trillion in profits, $13.7 trillion in revenues, $22.6 trillion in market value, and employment for nearly 30 million people around the world. If we are going to change the course of our economy, it is these very corporations that we must change. In the fight to save capitalism, this will be the battlefield.

PLAYING A NEW GAME

Monopoly's inventor, Elizabeth Magie, had a point to make. She wanted to illustrate the dangers of unfettered capitalism at a time when Gilded Age inequality roughly matched today's. Unlike the version we use today, Magie's original game in 1903 had two sets of rules. The first is the one we know: winning requires amassing a fortune by forcing others into bankruptcy. But in Magie's second iteration, the goal was different: instead of players bankrupting one another for individual benefit, they worked together to increase all of their wealth. The goal was to collaborate. Magie suggested playing the two versions in succession. That way, players would see the virtues of collective effort and the dangers of the alternative.

Ultimately, her hopes were dashed. After Parker Brothers bought the game in 1934, it marketed only the bankruptcy version. Parker Brothers thought it would sell better.

That's capitalism in the United States today—except with real money and real consequences. It's like the joke about the Planters mascot, Mr. Peanut: Is there anything more capitalist than a peanut with a top hat, cane, and monocle selling you other peanuts to eat? Corporations today act as if their sole and only purpose is to maximize short-term profits. The global pandemic revealed just how fragile our economy is. Never has it been more important to reimagine the corporation's role in society.

Through the United States' industrial period, paintings often depicted "Progress" as "a steam-powered locomotive, chugging across the continent, unstoppable." Today, we fear, capitalists are asleep at the switch, with dire consequences. When Benjamin Franklin sat at the Constitutional Convention, he often stared at the sun carved into the back of George Washington's chair. He would wonder to himself whether it depicted dawn or dusk. We might be forgiven for wondering the same thing about capitalism today. At the end of the successful convention, Franklin said, "Now at length, I have the happiness to know that it is a rising and not a setting sun." We hope the same for capitalism.

The legal scholar Lynn A. Stout once wrote, "We have been dosing our public corporations with the medicine of shareholder value thinking for at least two decades now. The patient seems, if anything, to be getting worse." We've played the first version of Monopoly long enough. It's time we tried the second.

1

CHALK ONE UP FOR THE GOOD GUYS

TWO CONFLICTING VISIONS FOR OUR ECONOMY

Guy Dixon's cousins never worked a day at the Kyanite Mining Corporation, but they owned a third of it. When they sued him over the generations-old family business, they did so as outsiders. The suit threatened to dissolve the company.

When people think of a socially responsible business, they don't usually think of a mine. But that's just what Dixon was trying to build: a mine that was both socially responsible and environmentally sustainable, though he doesn't use those words. For him, the company is rooted in an intergenerational mindset. He's committed to his workers. He's committed to his community.

"I'm doing what I'm doing for my kids," Dixon said late one fall afternoon, looking over the Virginia mine founded by his great-grandfather and then handed down to his grandfather, his father, and now him. The sun was starting to set behind the Blue Ridge Mountains off in the distance. "One day, hopefully, one of my kids will be doing what I'm doing." In the rest of the country, the average holding period of a stock is under a year.

A Dartmouth graduate and an alumnus of a white-shoe consulting firm, Dixon always knew he'd come back to his family's company, which mines the eponymous mineral kyanite in rural Virginia. Kyanite is used deep in the industrial economy, in things such as the lining of steel furnaces.

Dixon grew up working in the business. "You meet the people who are doing the real work," he said of his childhood. "You come to respect how hard, grimy, and dangerous their jobs are." He earned his right to run the company. He pulled up a video on his phone of his teenage daughter working a blowtorch to fix a bull gear. He scrolled through his photos until he found one of his son under a service truck changing a brake line.

Most companies rely on headhunters and external hires to fill the management ranks. A third of new CEOs are brought in from outside the company. But whereas the median job tenure in the United States is just over four years, at Kyanite it's close to twenty. All of Dixon's managers are homegrown. At Kyanite, it's not unusual for someone to retire at sixty-five—not sixty-five years of age but sixty-five years of employment. Dixon guessed that a third of the 160 employees had a parent or sibling who had worked there first.

The trust between the company and its workers has been built up over generations. As far back as Dixon can remember, the company has never fired anyone for financial reasons. When demand fell during the Great Recession and the economy was firing 700,000 workers a month, Dixon retrained miners to do other jobs rather than lay them off. "We couldn't just say, 'Okay, fellas, there's no work for you here.'" Instead, he found ways to keep his commitment.

What about the environment? Kyanite is a surface mine, which means that "liberating" the material—that's official terminology—requires removing all the soil and rock around it. Hanging in Dixon's office is a small black-and-white photo of a mountainside. Outside the window you can see the same mountain—except that half of it is now missing, blasted away with liquid dynamite, the kyanite itself crushed, purified, and transported to buyers worldwide. "Mining is a dirty business," Dixon admitted. "You make messes."

But unlike many mining companies, which are controlled by distant, anonymous shareholders, Dixon and his team are deeply ingrained in their community. "Our employees view the company as part of their family and their reputation," he said. And so when Kyanite shuttered its first mine, the company made a commitment to reclaiming the land back to a natural state. It won national awards for its environmental stewardship, beating global behemoths orders of magnitude larger. "We're this teeny, teeny Podunk business," said Dixon, "but all of us live around here—we use these woods and streams, we see and interact with our neighbors in the community all the time. That kind of drives us to do the right thing."

In their lawsuit, Dixon's cousins alleged that he and his father were running the company for their own benefit, cutting dividends and failing to maximize shareholder value, among other things. The case wound its way through the Virginia courts in the early 2000s. Fearing what might happen to the company if the Supreme Court of Virginia ruled against him, in 2013 Dixon settled with his cousins by buying them out.

Guy Dixon runs a surface mine. But he runs it in a way he is proud of. Whether that's enough depends, in part, on whether you think there can ever be such a thing as a socially responsible surface mine.

There are two conflicting visions in our economy today. One says that companies exist to make a profit and that managers should focus primarily on serving *shareholders*. The other says that companies should only earn a profit in pursuit of serving *stakeholders*—employees, customers, communities, and all those whom companies touch. We can see these conflicting visions in the Kraft Heinz Company's attempted takeover of Unilever in 2017. At Unilever, Paul Polman's leadership was based on a long-term stakeholder focus. At Kraft Heinz, Warren Buffett and the Brazilian private equity firm 3G

Capital had an explicit focus on shareholders. The takeover fight was—as one Harvard Business School case study framed it—a "battle for the soul of capitalism."

Though some business leaders like Polman endeavor to build prosperous companies that also serve some deeper purpose, most remain under the ongoing influence of Andrew Carnegie's "The Gospel of Wealth." Carnegie's book has us dividing our lives into two halves: economic in how we make our money, moral in how we then give it away. As corporations are increasingly pressured to appease both shareholders demanding profits and stakeholders demanding social responsibility, they are resorting to a rational hypocrisy: changing their marketing, not their mission.

In the end, corporations run for more than short-term profit can end up creating more long-term value for all stakeholders—shareholders included. But it's a hard road to get there. The same was true for Guy Dixon.

To keep Kyanite Mining Corporation operating after the settlement, Dixon sold off some assets and took on as much debt as the company could bear. That gave him cash to buy out his cousins' shares. When asked why he had cut the company's dividend—part of what had led to the lawsuit in the first place—he pointed to investments that have now, nearly fifteen years later, doubled the company's revenue. He tried to run his company with a long-term stakeholder focus because he *was* a long-term stakeholder—not just a shareholder.

Thinking about how it's all played out with his cousins, Dixon said, "I'd rather be in our shoes than theirs."

THE BATTLE FOR THE SOUL OF CAPITALISM

Paul Polman sat across the couch from his youngest son, Sebastian, for a taped interview. He crossed his leg casually over his knee. Between them were two mugs of tea—and a production worker holding a boom mic. It was part of #TalkToMe, a series of conversations between parents and children launched by the *Huffington Post* in 2016.

They talked about climate change, how society's expectations for businesses have changed over time, and how business can be a force for good. A few minutes in, they started to talk about what Paul Polman would do if he were no longer CEO. "You can always go back to your very first job ever, right?" Sebastian smiled.

"I could go back to my first job ever: a milkman in the Netherlands." They both laughed.

Just months later, Paul Polman—the stakeholder-focused, sustainability-proselytizing CEO of Unilever—took a meeting that he described as a "near-death" experience, when not only his career but all of Unilever would also be put at risk.

During the interview, though, he was at ease. He told his son about the important role milkmen play in society. When you're a milkman, a lot of people depend on you.

"Do milkmen still exist?" asked Sebastian.

"No, unfortunately not." Both Polmans laughed again.

Unilever, the $60 billion revenue consumer goods giant that owns brands from Dove soap to Lipton tea—and, yes, plenty of dairy products like Breyers ice cream—traces its history back to 1872. To many, Unilever is the poster child for social responsibility. Polman, who became CEO in 2009, built on the company's well-established social bona fides by putting sustainability at the center of strategy, setting long-term goals to serve customers in emerging markets, radically cutting emissions, increasing sourcing from smallholder farmers, and focusing on healthy and sustainable brands. As he put it, "When there are so many issues out there that need to be addressed, and when you are in a position to do something about that, I feel that sense of duty."

But all of that was suddenly at risk. One hundred forty-five years of corporate history could vanish over a few weeks, all starting with a lunch meeting over sandwiches at Unilever's London headquarters. The meeting was with the Brazilian private equity giant 3G Capital, cofounded by billionaires Alex Behring and Jorge Paulo Lemann. 3G had become infamous in the consumer goods market for its massive takeovers: Burger King, Budweiser, and the merger of Kraft Foods and Heinz. In some ways, Behring and Lemann were a strange pair of owners for those all-American brands. 3G controls a substantial portion of the world's beer production, yet Lemann doesn't drink. As *The Telegraph* of London described Lemann: "[He] would no more guzzle down a Magnum ice cream than glug Budweiser beer or chomp a Burger King Whopper. A lifetime of keeping fit—he was crowned Brazil's national tennis champion numerous times and played Wimbledon—has left the 77-year-old with a frame that matches the companies his investment firm 3G Capital controls: impressively lean."

Behring, meanwhile, says that "our greatest resource is our people." But 3G is notorious for cutting costs—including firing thousands of employees at the companies where Behring sits on the board. Its strategy is simple: buy big, bloated companies, and cut, cut, cut. It's known for a management technique called zero-based budgeting, in which every manager has to justify his or her budget from scratch every period. Every year—if not every quarter or every month—managers have to defend why each job should exist. If they can't, it won't.

Shortly after 3G acquired and merged Kraft and Heinz in 2015, it shuttered seven North American plants and cut a total of 2,600 employees. Two years later, it shut down the ninety-eight-year-old Oscar Mayer plant in Madison, Wisconsin, where four thousand workers had once made the iconic hot dogs. Kraft had had a troubled history even before 3G's ownership. When Kraft had acquired Cadbury in 2010, it had promised to retain factories and jobs in Great Britain. But it had reneged and closed a factory in Keynsham, reverting to Cadbury's earlier plan to move production to Poland. That was six hundred jobs now gone. Employees told the BBC that the move was "utterly despicable," "a cruel manipulation," and "a cynical ploy." One told reporters, "I believed Kraft totally, but they made a fool of me."

Within four years of 3G's takeover of Heinz, it laid off more than ten thousand people—a fifth of the combined workforce. Why? Because cutting costs increases profit, which increases share price, which is how 3G makes money. However, a focus on cutting costs hasn't helped Kraft Heinz boost its sales amid shifting consumer tastes toward healthier foods. In 2015, its UK division was heavily criticized for marketing its baby biscotti as "an ideal healthy snack." But British consumers weren't fooled, noting that the product contained 28 grams of sugar per serving. The company quickly amended its packaging after the complaint. Meanwhile, CSRHub (the letters stand for "corporate social responsibility"), which aggregates sustainability ratings on thousands of companies, rates Kraft Heinz in the fiftieth percentile. Unilever ranks near the top, in the ninety-fourth percentile.

When Behring went to Polman's offices that day, Polman thought he wanted to buy Unilever's waning "spreads" business—a portfolio of brands such as "I Can't Believe It's Not Butter!" that no longer fit Unilever's mission. But Behring wanted to acquire Unilever. All of it. He put forth an offer to acquire the entire company for $143 billion, an 18 percent premium to the current valuation in the stock market. 3G would combine Unilever with Kraft Heinz in the second largest deal in history. Polman was shocked. As he would later tell the *Financial Times*, the bid was "a clash . . . between a long term, sustainable business model for multiple stakeholders and a model that is entirely focused on shareholder primacy."

Unilever immediately pulled together a war council to defend itself: an expensive SWAT team of bankers, lawyers, and PR firms.

Polman—who had built his career on sustainability, long-term thinking, and putting customers first—must have been horrified by the idea of 3G taking control of Unilever. The corporation shut down the offer, writing in a public statement, "Unilever does not see the basis for any further discussions."

Behring had expected a warmer reception but was undeterred. Many of 3G's targets are reluctant at first before coming around at the prospect of a quick windfall for shareholders. These things often take time and sometimes a slightly higher offer. In its own public statement, Kraft Heinz wrote, "We look forward to working to reach agreement on the terms of a transaction."

The business world held its breath. Most corporate takeovers take months or even years to play out. As we'll soon see, this one would be over in days.

RESPONSIBLE CORPORATIONS, RESPONSIBLE CAPITALISM

Unilever itself is the result of a merger, but one nearly a century old. In 1929, the British Lever Brothers combined with the Dutch company Margarine Unie. Lever Brothers goes back to the 1890s, when William Hesketh Lever, running his family's grocery business in northern England, made it his mission to "make cleanliness commonplace; to lessen work for women; to foster health and contribute to personal attractiveness, that life may be more enjoyable and rewarding for the people who use our products." He would fulfill that mission with soap.

The result was one of the first branded personal hygiene products: a soap with the cheery name of Sunlight. Around the turn of the nineteenth century, Brits used less than four pounds of soap per person per year. By the end of the century, that had more than quadrupled to sixteen pounds.

The soap was manufactured in a large factory in Liverpool, where Lever built a village for workers with "a high standard of housing, amenities and leisure facilities." Unilever is proud of its social legacy: how during the Great War, the company agreed to reemploy all drafted employees on their return and even paid their dependents an allowance while they served; how it launched the Clean Hands Campaign in the 1920s to teach schoolchildren how to lather and rinse; how it expanded manufacturing in the 1950s in newly independent African states to serve customers in those markets. It goes on and on: sustainable fishing and no trans fats in the 1990s, hygiene education in rural India and sustainable palm oil in the 2000s, the Dove self-esteem project and zero waste to landfill since then.

When Paul Polman took the helm of Unilever, he had already spent twenty-six years at one of Unilever's biggest rivals, Procter & Gamble, and had served as CFO at Nestlé. At Unilever, though, he had a unique vision for leading the business into the future: to double the company's revenue while cutting its environmental impact in half. At the same time, he wanted to increase the share of revenue coming from emerging markets from less than half to over two-thirds. Just as William Hesketh Lever had focused his mission on those who would use his products, Polman believed, "My job is not to serve shareholders, but to serve Unilever's customers and consumers."

Polman stands out in that view, but he is not alone. Since the Great Recession, would-be capitalist reformers have begun offering new visions of what capitalism is capable of. One *Fortune* article described how "Harvard Business School professor Michael Porter began pushing what he called 'shared value' capitalism, and Whole Foods cofounder John Mackey propounded 'conscious capitalism.' Salesforce CEO Marc Benioff wrote a book on 'compassionate capitalism'; Lynn Forester de Rothschild, CEO of family investment company E.L. Rothschild, started organizing for 'inclusive capitalism'; and the free-enterprise-championing Conference Board research group sounded a call for 'sustaining capitalism.'"

As the journalist saw it, "Capitalism, it seemed, was desperately in need of a modifier."

Two oft-cited companies carrying the banner of social responsibility are Yvon Chouinard's Patagonia and John Mackey's Whole Foods. Both Patagonia and Whole Foods consider sustainability, responsibility, and accountability to be core to their business. As Chouinard put it, "Every time I have made a decision that is best for the planet, I have made money." Rose Marcario, the current CEO of Patagonia, recently updated the company's mission statement to read "We're in business to save our home planet." Mackey's book about Whole Foods is as good a playbook as exists for instilling and nurturing purpose in an organization.

Some have tried to turn this mentality into a movement, none more so than Certified B Corporations. B Corps are for-profit companies that explicitly seek to balance profit and purpose. After an evaluation, they are certified by the nonprofit B Lab, which was founded in 2006 by Andrew Kassoy, Jay Coen Gilbert, and Bart Houlahan. B Lab assesses the company's impact on everything from employees to products, from operations to its impact on the environment. According to B Lab, these companies "are accelerating a global culture shift to redefine success in business and build a more inclusive and sustainable economy." There are now more than 3,000 certified companies in over seventy countries. Though these include major brands such as Seventh Generation and Ben & Jerry's—both subsidiaries of Unilever—most B Corps are smaller or younger companies, many still led by mission-driven founders.

Cofounder Andrew Kassoy believes B Corps are important leaders in the push to change capitalism. "The theory of change," Kassoy said, "was to build a community of credible leaders who have met very high standards and opted in." These were the early B Corps, whose social purpose and commitment were core to their identities. They would both inspire others and provide them with the tools they needed. "Until we built a community of these credible leaders," he said, "we weren't going to get public companies with their public shareholders to join." Though few large-scale enterprises have joined the movement—of the 3,700 or so public companies in the United States, only three are B Corps—Kassoy believes it is now hitting an inflection point.

Danone North America became the world's largest Certified B Corporation in 2018. In 2019, the Gap announced plans to become one as well, something its subsidiary Athleta had achieved the year before. The Body Shop also became a B Corp in 2019 as a subsidiary of Natura & Co, a publicly traded B Corp. The debate around corporate social responsibility and corporate purpose is now reaching a fever pitch—fifteen years after Kassoy helped found B Lab. Does it feel like he's winning the war to reform capitalism? "On the one hand: yes, I think the last year has been a huge victory, and I think we've played our little part in helping to lead to that," he said. "On the other hand, it sets up a massive challenge for what the next ten years of our work will look like." It's no longer enough to build up a community of B Corps. They now have to use that community to change the way all business is done.

Just because a company is not a B Corp does not mean it cannot be run according to similar principles. Part of the B Corp movement's success will be measured in how far its ideals and practices extend beyond the confines of its own community.

American corporate law is clear that managers must act in the best long-term interest of shareholders. But courts generally give managers broad discretion, including the decision to follow the sort of inclusive principles that B Corps embody. This, known as the business judgment rule, allows managers to do things such as increase pay or invest in the local community even if it hurts current profits. The clearest example is corporate philanthropy, in which managers can donate money so long as they believe it will create goodwill for the company, bolster its reputation, or in some other way benefit the company and its shareholders over the long term.

Though Unilever is not a B Corp, Polman speaks as if it were. "Business is here to serve society," he wrote for the *McKinsey Quarterly* in 2014. And: "Why should the citizens of this world keep companies around whose sole purpose is the enrichment of a few people?"

FOR GOD AND FOR PROFIT

3G's intentions with Unilever were based on two numbers: 23 and 15. The difference between them could be worth billions of dollars.

Those were the operating profit margins of Kraft Heinz under 3G's ownership and Unilever under Polman's. For every dollar of revenue that Kraft Heinz made, 23 cents was profit. A dollar of revenue at Unilever made only 15 cents. There could have been many reasons for this difference—different products, different geographies, different strategies—but Behring was betting on one: inefficiency. Unilever's CSR focus might have been holding it back, adding unnecessary expense to its operations. If it cut costs, raised prices, and focused on its most profitable products and customers—rather than, for example, the neediest—it could increase its profits. For investors such as 3G, improving efficiency is the core premise of capitalism.

It was not alone in this assessment; 3G's bid was backed by one of the most legendary investors in US history: Warren Buffett. Buffett had partnered with Behring and Lemann at 3G to buy Heinz for $28 billion in 2013. "This is my kind of deal and my kind of partner," he said at the time. It was an iconic brand with a durable competitive position. What could beat Heinz ketchup? Together they combined it into a $45 billion merger with Kraft Foods two years later.

The cost-cutting private equity titans of 3G were not such odd partners for the folksy, avuncular Warren Buffett. More than anything, Buffett admires sustained growth in productivity. In his 2015 letter to shareholders, he gushed that 3G "could not be better partners." He continued, "Their method, at which they have been extraordinarily successful, is to buy companies that offer an opportunity for eliminating many unnecessary costs and then—very promptly—to make the moves that will get the job done. Their actions significantly boost productivity, the all-important factor in America's economic growth over the past 240 years."

For Buffett, increasing productivity is not just a way to make money. He believes that by maximizing profit, corporations become drivers of progress in our country and our world. At Buffett's annual shareholder meeting in 2019, an event some call "Woodstock for Capitalists," he said that the goal of capitalism is "to be more productive all the time, which means turning out the same number of goods with fewer people or churning out more goods, with the same number." Through this competition for profit, the people and planet will be well served.

One problem, of course, is what Buffett himself called "roadkill": the people displaced as capitalism uses labor more efficiently. For Buffett, that problem is mostly for government to solve. He sees the growing corporate social responsibility movement as largely virtue signaling and ultimately counterproductive. The legendary venture capitalist Marc Andreessen took a similar stance in his assessment of B Corps. "I would run screaming from a B Corp," he said. "The split model makes me nervous and I don't think we would ever touch that. It's like a houseboat. It's not a great house and not a great boat." As Buffett's longtime partner Charlie Munger summed it up, "I like our way of doing things better than theirs and I hope to God we never follow their best practices."

In 1970, the Nobel Prize–winning economist Milton Friedman made a compelling case for this sort of economic thinking in the *New York Times Magazine.* Its title has emboldened investors and enraged activists for decades: "The Social Responsibility of Business Is to Increase Its Profits."

Friedman began by arguing that businesses cannot have responsibilities; only people can. Corporate *managers* are people, but they are first and foremost employees of the shareholders who own the corporation and so must be responsible to *them*. Shareholders own stock because they want to maximize their profit. Therefore, if corporate managers are responsible to shareholders who just want to maximize their profit, then corporate managers should make as much money as possible.

Friedman was one of the most effective public intellectuals of the twentieth century because of his ability to simplify complex topics in a way that rings true. It makes sense to think of the CEO of General Electric as working for the shareholders of General Electric, who collectively own the company. And it makes sense to think that shareholders want to maximize their profit. Why else would you invest in a stock if not to make money?

He's transformed a morally rich question into a tautology: if you're hired to do a job, your responsibility is to do it. If that means taking actions that harm employees, communities, the environment, or even customers, so be it. Managers are not meant to be moralizers. They should just do their job. Friedman's economic system represents an adversarial division of labor. Corporations maximize profit. Government keeps them in line. Environmentalists protest. Buyers beware. Workers take cover. Nonprofits pick up the pieces.

The hope is that out of these conflicting roles we can forge a prosperous and just society. This is not Polman's vision of corporations in society, but it is Friedman's. And it was this sort of thinking that put Polman's vision into jeopardy.

"THE GOSPEL OF WEALTH"

Buffett and his ilk are not heartless. Far from it. In 2006, Buffett pledged to donate more than 99 percent of his wealth to charity.

"My wealth has come from a combination of living in America, some lucky genes and compound interest," he wrote. "My luck was accentuated by my living in a market system that sometimes produces distorted results, though overall it serves our country well." Buffett launched the Giving Pledge with Bill and Melinda Gates in 2010. It's a commitment by the world's wealthiest individuals to give away at least half of their money. By 2019, it had garnered 204 signatories from twenty-three countries.

Though the Giving Pledge is new, the idea is not. In fact, it can be traced to a different capitalist of an earlier era who similarly bifurcated his life into first making money and then giving it away: Andrew Carnegie.

For most of Andrew Carnegie's career, he was feared as ruthless. Trusts such as his U.S. Steel, which combined numerous former competitors into a cartel with the power to dictate prices, were what animated the trust-busting presidency of Theodore Roosevelt. U.S. Steel employed 250,000 people at its height, more than the country's armed forces.

Carnegie squeezed his employees and customers alike, sometimes using brutal methods; sixteen people died in a single day when he used a private army to break his striking workers in Homestead, Pennsylvania.

In the process of building his empire, he amassed a fortune of nearly $100 billion in today's dollars. But then he decided to give it all away. "The man who dies thus rich dies disgraced," he wrote. His manifesto on philanthropy, "The Gospel of Wealth," has since become sacrosanct in some circles of charitable giving. It includes some interesting proposals for moderating capitalist excess, including steep inheritance taxes: "By taxing estates heavily at death," he wrote, "the state marks its condemnation of the selfish millionaire's unworthy life." The Philanthropy Roundtable declared him to be perhaps the most influential philanthropist in American history.

Carnegie laid out an apparently virtuous two-step: First, make as much money as humanly possible by any means possible. Second, give it all away for the betterment of mankind. He endowed 2,811 libraries, bought 7,689 organs for use in churches, founded one of the world's great research universities, and set up numerous charities that are still operating today. His works represent a set of dueling impulses: those of a titan and a philanthropist both. Fatal accidents in steel mills of the 1880s accounted for 20 percent of all male deaths in Pittsburgh, and Carnegie the Titan did little in his capacity as the owner of U.S. Steel to improve safety standards. But then he donated the money made on the backs of those very workers to support the injured. His first gift was a $4 million fund for workers as an "acknowledgement of the deep debt which I owe to the workmen who have contributed so greatly to my success." "The Gospel of Wealth" allowed him to do both. It was Carnegie the Titan's single-minded pursuit of profit that begot the good works of Carnegie the Philanthropist.

In the United States, this philosophy helps explain a giving rate that is much higher than that of most other countries. For decades, Americans have consistently donated the equivalent of 2 percent of GDP to charity each year. Our worry is that such giving is rendered impotent without the other 98 percent of GDP pushing in the same direction.

This tension between doing well and doing good lives in each of us. An informal poll of Stanford business students revealed that 95 percent of them felt some tension between doing well in their careers and doing good for society. Carnegie's example demonstrates this tension: We are both moral and economic beings, and our current system often pits these competing impulses against each other. We have become accustomed to living with that friction, along with a kind of background guilt that characterizes the countercurrents of our nature. "There is something within all of us," preached Martin Luther King, Jr., "that causes us to cry out with Plato that the human personality is like a charioteer with two headstrong horses, each wanting to go in different directions." Not only is there a gulf between what we profess and what we practice, there's often a contradiction between what we do to make money and what we do to make good. Said King: "There is that persistent schizophrenia which leaves so many of us tragically divided against ourselves."

The hedge fund manager Paul Tudor Jones II described it this way: "We have put so much emphasis on profits, on short-term quarterly earnings and share prices, at the exclusion of all else. It's like we've ripped the humanity out of our companies." Together with other socially minded leaders, Jones created the nonprofit JUST Capital, which "tracks, analyzes, and engages with large corporations and their investors on how they perform on the public's priorities." With rankings, polls, and other data-driven tools, JUST Capital hopes to help both companies and shareholders bring that humanity back to the business world.

Christopher M. James, another hedge fund manager, was raised closer to Carnegie's ideology. He grew up in southern Illinois the son of a coal miner, reading biographies of Andrew Carnegie and John D. Rockefeller and dreaming of moving to New York City to follow in their footsteps. But decades later, as he was making his own fortune as an investor, he realized that he was living at odds with his values.

In his personal life, he devoted his time and money to the Theodore Roosevelt Conservation Partnership and the National Fish and Wildlife Foundation. He was funding lawsuits to prevent drilling for oil in national forests. But in his professional life, he was investing in oil terminals. In his personal life, he founded the nonprofit Tipping Point, which fought homelessness in San Francisco. In his professional life, he was investing in the very tech companies that were contributing to the city's housing crisis.

While James was visiting a friend's family home in Virginia, he saw a painting of a Confederate colonel. The colonel, his friend's ancestor, had obviously once been a point of family pride. But times change, and moral standards evolve. James found himself staring at the painting. "And I had this thought," he said. "Are my kids going to say, 'My dad made his money in tech companies that destroyed San Francisco and oil companies that destroyed the environment?'"

So he stopped. "I stopped doing any investing, public or private, that was not consistent with my values." He shut down his hedge fund, returning billions of dollars to shocked investors. Following "The Gospel of Wealth" had divided James against himself. He committed to spending a year figuring out how to devote all of his resources—personal and professional—to fight for what he believes in. His own journey is just beginning. To convince others, he knows, he'll have to lead by example.

Capitalism today tries to resolve this paradox of wanting to do well and do good by dismissing one side of our nature or the other. Economists talk of efficient markets, as if economic activity were bound by eternal laws. And so we bring cold hearts to the marketplace, which is then cold to us in return. Philanthropical foundations pull at our heartstrings and so we give back some of what we've earned.

The result is the contradiction at the heart of "The Gospel of Wealth." The same owner allows men to die in his mills and then donates to a charity that helps the injured. In his assessment of Carnegie's philanthropic acts, Theodore Roosevelt wrote, "If Andrew Carnegie had employed his fortune and his time in doing justice to the steelworkers who gave him his fortune, he would have accomplished a thousand times what he has accomplished."

Then there are those such as Jack Bogle, the late founder of the $4.9 trillion asset manager Vanguard Group. His peers boast wings at museums, graduate schools at universities, and hospitals in their names. In the Carnegie tradition, Stephen Schwarzman—

the founder of the asset manager Blackstone Group—underwrote a $100 million restoration of the New York Public Library. Its main branch was renamed in his honor.

Bogle took a different approach. He set up Vanguard as a "mutual company," meaning it is owned by its customers—the investors in its funds. When Vanguard turns a profit, it reinvests that money into lowering costs for customers. This allows it to charge fees that are 83 percent below those of its competition. "In investing," Bogle said, "you get what you *don't* pay for." From his perspective, this was the purpose of a financial company: help your investors save for retirement, taking just enough to cover your expenses and make a fair income. But that meant that he would never amass the billion-dollar fortunes of his competitors. When Bogle passed away in 2019, his net worth was 99 percent less than the $7.4 billion of his rival Edward Johnson, the chairman of Fidelity Investments, even though Vanguard has more assets under management.

Few outside the financial world know Bogle's name. But every year, he saves investors $100 billion, money that goes directly to their retirement funds, college tuitions, down payments, and savings. He did not make billions of dollars charging investors hidden fees on complex financial products they did not understand, only to give it back to society in the form of a building with his name on it. He did not bifurcate his life into separate economic and moral selves. He rejected "The Gospel of Wealth," saving his customers money instead.

"This whole idea around the Giving Pledge is a kind of lazy man's way to do this," said Chris James. "If [Mark] Zuckerberg spent more time making Instagram and Facebook actually good for our kids rather than addicting our kids, he'd have a much bigger impact." In James's view, philanthropy will always lack the scale to make the changes we need if it's working at cross-purposes with corporations, consumers, and capitalism. "When you think about this," he reflected, "it's just so obvious."

RATIONAL HYPOCRISY AND THE TWO-BUFFETT PARADOX

For one Columbia professor, there is no question anymore about whether CEOs should be advancing society's goals. "Profit and purpose are converging," he wrote in *Harvard Business Review*, rejecting "The Gospel of Wealth." Born in 1983, he is of the generation that no longer believes the only purpose of business should be to make a profit. That professor? Warren Buffett's own grandson, Howard W. Buffett.

With coauthor William B. Eimicke, Buffett argued, in *Social Value Investing: A Management Framework for Effective Partnerships*, that corporations need to assess how they are affecting society and change accordingly. It's the vision espoused by Polman in his defense of Unilever, it's the vision embodied by B Corps, and it's the vision that younger generations like Howard Buffett's are increasingly adopting. In one survey, 40 percent of millennials said they'd taken jobs because of companies' sustainability performance, while only 17 percent of baby boomers said the same. In another survey, 62 percent of Gen Z consumers said they preferred sustainable brands, while only 39 percent of baby boomers agreed. As employees and consumers, younger generations are increasingly demanding that corporations act in the best interests of society and the environment.

Just one problem: corporations still have to answer to shareholders.

The two Buffetts offer conflicting visions of the corporation. Buffett the elder is skeptical of corporate social responsibility, taking the traditional view that corporations should focus on maximizing shareholder value. Buffett the younger believes corporations should focus on creating social value.

Today, corporations increasingly face two contradictory demands: "Maximize profit," we tell them. "But benefit society!" That's all well and good when the two clearly line up. But what about when the trade-offs between short-term profit and broader conceptions of corporate value are muddy? Some describe this as the difference between shareholder capitalism and stakeholder capitalism. It's not just an academic debate—at least, not for the corporations that are being forced to answer to both demands.

This is the Two-Buffett Paradox: business leaders are simultaneously asked to lead according to the conflicting worldviews of both Buffetts. For many corporations, the rational response is hypocrisy: with different types of stakeholders voicing seemingly irreconcilable demands, it makes sense for corporations to say different things to different audiences and then continue to serve the priorities of shareholders above all. This enables managers to pacify reformers without seeming to sacrifice investors. It's easier to fake good works than good returns.

Even in cases where corporations do benefit society, they justify it according to the bottom line. It's not that a corporation *wants* to invest in the local community. It's that it *needs* to invest in the local community in order to maximize its profits. Within the Milton Friedman framework, corporate social responsibility can exist only as a means to an end, not as an end in itself. It's tolerated only when it's insincere.

Even some of the most ardent social activists have adopted Friedman's logic. They repeat the mantra of "doing well by doing good." They passionately advocate for the environment, the poor, or our communities. Then they sort of wink and say, "Oh, by the way, this is also how to make the most money." What's not to like?

Unfortunately, when socially conscious corporate managers justify CSR with profit, they are yielding ground to the same short-term, profit-obsessed worldview that they seek to challenge. It's a hidden irony lurking within reform circles today: the CSR-focused manager believes in the primacy of profit no less than did Milton Friedman himself. If it turns out that caring for the environment or increasing diversity does *not* maximize short-term profit, activists are stuck either changing their principles or changing their logic.

The conflicting demands of the Two-Buffett Paradox are forced on all public companies, Unilever included. Though Polman enjoyed broad support for his sustainability initiatives, he still felt the pressure to increase profitability.

When Polman took the helm at Unilever in 2009, his stated goal was that the company would endure for another century. In public companies, managers usually report to investors on their financial performance once every quarter. They also often give projections for the quarter to come. Polman ended that practice at Unilever, saying, "We needed to remove the temptation to work only toward the next set of numbers."

Polman's leadership was based on the idea that profitability must be put into a larger context. He was trying to demonstrate that *why* you do something can have a profound impact on *how* you do it and in what will result. Two identical actions, one explicitly justified by profit and the other by a moral responsibility, could have two very different outcomes. It's like showing love to your children: only when you do it unconditionally without the expectation of rewards are rewards likely to come.

Against the Two-Buffett Paradox and the rational hypocrisy it often elicits, Polman tried to resolve the tension by advancing a long-term agenda, characterized by an interest in stakeholders and the environment for their own sake. He believed that doing so would also serve shareholders well, but, critically, that was not his primary focus.

Polman retired in 2019. To the extent that he proved himself to be an enlightened leader, it will be for balancing the competing demands he faced in a way that served all his stakeholders well.

CLOSING THE DEAL

As news of Kraft Heinz's potential takeover of Unilever spread, the British government saw it as a matter of public concern. The news of the takeover broke on a Friday, February 19, 2017. Over the weekend, Prime Minister Theresa May and other leaders began raising questions. For Kraft Heinz to be successful, it would have to convince more than just the Unilever board. It would have to convince the British public, too, or risk a government block of the transaction.

As it would turn out, Kraft Heinz would never have the chance. At 5:31 p.m. Sunday, just over a week after Alex Behring approached Paul Polman with 3G's plan to merge Kraft Heinz with Unilever, the latter two companies issued a joint press release. Earlier that day, Behring, Lemann, and Buffett had received a letter from Unilever. Their offer had "no merit, either financial or strategic," it wrote. If they wanted to proceed, they would have a fight on their hands.

What happened next is unclear. With so much at stake, many expected 3G to push forward. Unilever's share price had been mired between $40 and $45 for years. Kraft Heinz was offering $50. Surely the offer would have been enticing to many shareholders, eager for a quick return on their investment. But the joint press release ended all speculation. "Unilever and Kraft Heinz hereby announce that Kraft Heinz has amicably agreed to withdraw its proposal for a combination of the two companies. Unilever and Kraft Heinz hold each other in high regard. Kraft Heinz has the utmost respect for the culture, strategy and leadership of Unilever." With that, the entire episode ended almost as abruptly as it began.

News reports suggested that Buffett and Lemann wanted "to avoid a potentially dirty and public takeover battle." One hedge fund manager remembered being with Polman the week of the bid. "I've been here before, Paul," he told Polman. "Play your fucking card with Warren because he does not want you to talk to the world about the shit Kraft Heinz does to its consumers." In Polman's assessment, Buffett wasn't actively involved in the initial bid, and when he discovered it would become a hostile one, he pulled back.

Buffett and his partners at 3G wanted to own Unilever, to combine it with Kraft Heinz, to cut costs and increase efficiency across the storied business. They just didn't want to have a public fight about it. The *Financial Times* called it a humiliating reversal: $143 billion did not change hands. Unilever remained an independent company. Paul Polman and his unique view on sustainability and strategy lived to see another day.

Business leaders such as Paul Polman at Unilever, Yvon Chouinard at Patagonia, and John Mackey at Whole Foods show us that it's possible to build a better capitalism through better corporations—corporations that serve all their stakeholders over the long term. Yet they remain lonely pioneers on a frontier otherwise dominated by Milton Friedman acolytes. It's a frontier where the Two-Buffett Paradox reigns and rational hypocrisy is still too often the standard attitude.

The terrible and wonderful thing about a public company is that its stock price is always available. It's an erratic measure, but it's a way those in business and finance obsessively keep score. One of Warren Buffett's professors at Columbia Business School in the 1950s coined a famous description of the stock market that's still quoted frequently today: "In the short run, the market is a voting machine . . . but in the long run, the market is a weighing machine."

At any given time, stock prices reflect hype cycles and the madness of crowds. The tendency for share prices to swing wildly is captured by a joke about the infamously volatile cryptocurrency Bitcoin: A son asks his dad for $10 of Bitcoin. The dad replies, "$8.50? What do you need $11.25 for?" Over the long term, however, business fundamentals tend to win out. At first, perception is reality. Ultimately, reality is reality.

In the two years following Kraft Heinz's failed takeover of Unilever, Unilever's share price has grown 40 percent to nearly $60, well above the $50-per-share offer it had received. Over that same period, Kraft Heinz lost an astonishing 61 percent of its value, tumbling in price from $70 per share to $27. Its CEO has stepped down, and the Securities and Exchange Commission is investigating its accounting practices. A Reuters columnist argued that a takeover now made sense, but that it should be Unilever acquiring Kraft Heinz.

After Kraft Heinz's decline, Warren Buffett went on CNBC to defend his investment. He sat there, split screen with a red graphic of the share price decline above a breaking news quote: "Buffett: I was wrong in a couple of ways on Kraft Heinz." He admitted that he had overpaid. He admitted that the company had struggled to adapt to changing customer preferences for healthier, more sustainable food.

Both Kraft Heinz and Unilever are companies with more than a century of history. Their long-term value will be determined less by the costs they cut next quarter than by their customers, their employees, and their communities. And for the moment, at least, the market is voting for Polman.

Now Kraft Heinz is trying to catch up. A year after its failed bid for Unilever, it released its first corporate social responsibility report. In it, it committed to transitioning toward 100 percent cage-free eggs, to reduce energy and waste by 15 percent, and to deliver a billion nutritious meals to people in need. It's all in line with its new mission "to be the best food company, growing a better world."

Is Kraft Heinz the latest convert to stakeholder capitalism? Or is its new mission just a marketing ploy? The fight to save capitalism exists in this tension between idealism and cynicism. We want to believe that corporations are transforming themselves, but we're skeptical that they actually are. Kraft Heinz changed its mission. Even if every other corporation followed suit, would that be enough to change the system?

"In my mind, system change requires three things," said Andrew Kassoy, the cofounder of B Lab. "It requires behavior change, it requires culture shift, and it requires structural change. And those things all feed on each other." All the talk about corporate social responsibility and corporate purpose reflects a significant shift in culture. But that's not enough.

We have built our economy to get a single outcome: maximizing shareholder value. It's not that people are evil or that mission statements are necessarily hollow. It's that corporate managers and investors remain accountable to nothing but profit. "The only way you really hold corporations' feet to the fire," said Kassoy, "is if they're actually accountable to do it. And that's the part I'm cynical about."

UNDERSTANDING THE SYSTEM

In some ways, it is surprising that Unilever and Kraft Heinz are as different as they are. After all, both are global conglomerates subject to the same forces of modern capitalism: the stakeholder demands to be good and the shareholder demands to be profitable. Both are subject to the Two-Buffett Paradox. Given this, we need to understand why Unilever is the exception rather than the rule. We need to understand why so many corporations are run for short-term profit rather than in the long-term interest of all stakeholders. We need to understand why so few companies reflect our values.

The key may lie in an unexpected place: a small mining operation in rural Virginia. If we're going to drive cars or ride railways or make things with industrial ovens, someone needs to mine kyanite. And if someone needs to mine kyanite, better for it to be Guy Dixon than someone who thinks primarily in terms of quarterly results. But the things that make Dixon a responsible business leader are mostly absent in public companies today. In the next chapter, we'll see how local, intimate ownership has given way to distant, anonymous shareholders. We'll see how generational thinking has given way to an obsession with quarterly earnings. We'll see how implicit contracts between stakeholders built on trust have given way to a dog-eat-dog ethos of adversaries.

Dixon's cousins never worked at the mine they partially owned, just as few of today's shareholders have ever worked at the companies they're invested in. Few could even *name* the companies they own through their mutual funds and retirement accounts. For most corporations today, a rich web of duties, values, and mutual obligations has been reduced to a single goal: maximizing share price. That's fiduciary absolutism. And that's where we will go next.

2

MAKING A KILLING

WHY FOOD COMPANIES DON'T NOURISH US

When Curt Ellis graduated from college, he gave researchers a single strand of hair. After examining its molecular makeup, they told him the composition of his diet. Like that of most Americans, it was based overwhelmingly on corn. Ellis was surprised; he didn't think he ate much corn.

Walk down any supermarket aisle, and you will find corn processed into every imaginable form. Bread, juice, granola bars, yogurt—read the labels, and there it is. Your salad at lunchtime? Drenched in salad dressing sweetened with high-fructose corn syrup. The beef you eat? Fattened on corn. If we are what we eat, it's a wonder we haven't all turned yellow.

Shocked by that discovery, Ellis and a classmate decided to keep digging. After college, they moved from Connecticut to Iowa, where nearly a fifth of the nation's corn is grown. With a population of 3 million people, Iowa grows nearly 2.5 billion bushels of corn annually. Curt and his partner bought an acre of land to grow corn of their own. Their plan was to trace its journey through the US food system.

The result was *King Corn*, a documentary that serves as a scathing indictment of our food industrial complex. Like Upton Sinclair's *The Jungle*, which revealed the horrors of Chicago slaughterhouses in the early 1900s, *King Corn* pulled back the curtain on the modern American diet. It's a story of government subsidies, corporate greed, and an ambivalence about nutrition that have conspired to help make Americans among the least healthy people in the developed world.

For parents with limited time and small budgets, the easiest thing to do is feed kids empty, corn-based calories. Our food system enables our least nutritious and most caloric cravings, engineered and reengineered to stuff shareholders' pocketbooks and consumers' expanding waistlines. As a result, one in five children today is obese—a rate that has tripled since the 1970s. For children of color, it's one in two, as is the probability of developing type 2 diabetes at some point in their lives. Forty percent of American adults are obese, resulting in conditions ranging from diabetes to stroke, cancer, and heart disease. Poor diet is now the leading cause of death around the globe.

No farmers are *trying* to sicken their consumers, just as no financial advisers are trying to impoverish their clients, no educators are trying to saddle students with debt but no degree, and no doctors are trying to leave patients addicted to opioids. No one intends it. But it happens.

The field of design thinking has a saying: "Every system is perfectly designed to get the results it gets." This chapter is about understanding how our system has given us the corporations we've got. We will begin with the story of Kellogg's. Kellogg's was founded to make breakfast healthier but eventually succumbed to selling sugary cereals instead. We will see how this reflects a broader disease in our corporations: fiduciary absolutism, a myopic focus on short-term profit above all else.

We will explore the theoretical underpinnings of fiduciary absolutism. It began as an answer to a question that is long out of date: How can we make corporate managers *more* focused on serving shareholders? We will see how fiduciary absolutism relies on economic theories—the efficient market hypothesis and the invisible hand—that apply better to the eighteenth-century Scottish economist Adam Smith's age than to our own.

But where did fiduciary absolutism come from? In this chapter, we will also trace a brief history of capitalism in America. Ultimately, we will see how the economic malaise of the 1970s combined with structural changes to our economy to create fertile ground for fiduciary absolutism to take root.

That does not mean our corporations were perfect in the past, nor that they are damned to chase short-term profits for all time. We'll see how some new companies are challenging fiduciary absolutism and how larger corporations—including Kellogg's itself—are now trying to adapt to a changing world.

Food companies are being pushed along by people such as Curt Ellis. Not content with just making a documentary, Ellis began to fight. In 2010, he cofounded FoodCorps, which embeds recent college grads in high-poverty schools to teach healthy eating. By the time many students get to school in the morning, they have already consumed 15 or more grams of sugar in a bowl of cereal such as Frosted Flakes. That's 60 percent of the recommended daily limit even before first period has begun.

Ellis is fighting for the health of our children, but he's up against the multitrillion-dollar food industry. How can he combat the 225 million hamburgers McDonald's sells worldwide each year? Or the equivalent of 2 billion cans of carbonated sugar water Coca-Cola sells *each day*? Ellis could spend his life preaching the virtues of fresh fruits and vegetables, but when Taco Bell makes a taco shell out of Doritos, it sells a million tacos in twenty-four hours.

Five of the largest food manufacturers in America—Kellogg's, General Mills, Tyson Foods, PepsiCo, and Kraft Heinz—sell over $150 billion of product each year. Reformers like Ellis aren't sticking a finger into a dike; they're caught in a tidal wave.

Despite Big Food's cynical marketing to our worst biological impulses, the tale is not all about unadulterated corporate greed—at least not from the beginning. In fact, some of the worst corporate actors today were founded with noble ambitions. And it all started with breakfast.

THE ORIGINAL SUPERFOOD

Kellogg's Corn Flakes was meant to be a superfood. It was developed by a doctor for his patients as an easy, healthy substitute for what passed for breakfast at the time. During the mid–nineteenth century, the word "dyspepsia" entered the American lexicon. Howard Markel, the author of *The Kelloggs: The Battling Brothers of Battle Creek*, described dyspepsia as "a nineteenth-century catchall term for a medley of flatulence, constipation, diarrhea, heartburn, and 'upset stomach.'" He went on to show that breakfast was one of the main contributors to the condition:

> Early morning repasts included filling, starchy potatoes, fried in the congealed fat from last night's dinner. As a source of protein, cured and heavily salted meats, such as ham or bacon, were fried up as well. . . . the staggeringly high salt content made one quite thirsty and eager for a drink—a situation not lost on the saloonkeepers of every town in America who routinely opened for business in the morning.

Enter Dr. John Harvey Kellogg of Battle Creek, Michigan, and his brother Will. Like Curt Ellis, they were men on a mission to improve the American diet. At the Battle Creek Sanitarium, the Kellogg brothers experimented with new ways to feed patients that focused on whole grains and fresh vegetables and fruits. As one popular myth has it, Kellogg's Corn Flakes came to John in a dream involving a new way of rolling and cooking a batch of wheat-berry dough. The ultimate result was a simple and—for its time—healthy dish that revolutionized the food industry.

Few companies can claim to have changed the way Americans eat, but Kellogg's and its early competitors did just that with breakfast cereals. Unfortunately, Kellogg's Corn Flakes and the other better-for-you brands that launched this industry have been overcome by the countless sugary cereals that line today's supermarket aisles. Roughly 90 percent of Americans indulge in packaged cereal of some kind, and nearly all of these cereals are based substantially on added sugars.

The cereal Cookie Crisp was marketed to kids during cartoon commercial breaks with the tagline "You can't have cookies for breakfast. But you *can* have Cookie Crisp!" Similar cereals are often more than one-third sugar by weight. That's like eating a spoonful of sugar for every two spoonfuls of Kellogg's original cereal. Cookie Crisp may wear its disregard for health on its sleeve, but Kellogg's Honey Smacks touts misleading nutrition claims such as "Good source of Vitamin D" despite providing only a tenth of the recommended daily value. Raisin Bran sounds healthy, but it has more sugar per serving than Count Chocula.

How did Kellogg's go from manufacturing the first breakfast superfood to becoming a supervillain of the American diet? Dr. John Harvey Kellogg wanted Kellogg's to help sick people get healthy. What would he make of a world in which his name has become synonymous with making healthy people sick? Though Kellogg's was founded with a deeper purpose, it is representative of many of our large corporations today: it appears to care more about earning profits for shareholders than it does about the long-term well-being of its customers—not to mention its employees or its communities. This ideology—which we'll call fiduciary absolutism—has become so pervasive that we accept it as just the way the world works. As we'll see, the forces behind Kellogg's transformation are pushing on every corporation in our economy. But when we look at the company's history, we see that the world hasn't always worked this way, and it need not in the future.

THE THEORIES UNDERLYING FIDUCIARY ABSOLUTISM

In 1976, the bicentennial of our nation's independence and the 200th anniversary of the publication of Adam Smith's *The Wealth of Nations*, the economists Michael C. Jensen and William H. Meckling published a paper that has since become the most widely cited study in the history of business literature. It's done as much to influence the way our economy works as any other academic paper in history. Its subject? The principal-agent problem.

Imagine this: You hire a babysitter to watch your children. You are the principal, and the babysitter is your agent. The problem is that your interests diverge. You care much more about your children than your babysitter does. Your babysitter cares, sure, but he also cares about having an easy afternoon. When your children act up, it's easier for the babysitter to put on the TV than provide a lesson on sharing. Since the agent has the day-to-day control, the principal's interests tend to suffer. This is the principal-agent problem.

Jensen and Meckling argued that this was *the* central problem of corporate governance. Corporate managers (the agents) were not working in the best interest of shareholders (the principals).

To solve this problem, they suggested clarifying what a corporation is for. "Because it is logically impossible to maximize in more than one dimension," Jensen argued, "purposeful behavior requires a single-valued objective function." Under the principal-agent problem's framework, the *single-valued objective function* should be for the corporation's managers to *maximize financial value for shareholders.*

This theory built on the existing law of fiduciary duty. Fiduciary duty is as old as the Code of Hammurabi. Some version of it is law everywhere that corporations exist. Fiduciary duty mandates that managers not be self-serving or negligent in running corporations owned by others. It provides some guardrails on what managers can do. For example, managers cannot sell corporate assets to themselves at discounted prices.

Fiduciary duty is a vital part of corporate law. But in their paper, Jensen and Meckling proposed something radical: they took the guardrails of fiduciary duty and transformed them into the sole purpose of the corporation. In other words, fiduciary duty became fiduciary absolutism, and a constraint became a demand. With support from popular intellectuals such as Milton Friedman, fiduciary absolutism became the guiding principle of the modern US economy. Forget balancing different stakeholder demands or conflicting obligations to society; just maximize shareholder value. As the legal scholar Lynn Stout wrote, this "offered an easy-to-explain, sound-bite description of what corporations are and what they are supposed to do."

The idea caught on. As one contemporary remembered it, Jensen's arrival at Harvard Business School, in 1985, represented a singular shift. Before his arrival, "No one was talking about 'shareholder value.'" But by 1986, after Jensen was ensconced as a tenured professor at the nation's most prestigious business school, everyone was. Business research since the 1970s has been focused almost entirely on financial outcomes, rather than corporations' many other roles in society. At that point, Stout wrote, "Shareholder primacy had become dogma, a belief system that was seldom questioned, rarely justified, and so commonplace most of its followers could not even recall where they had first learned of it."

We can trace the spread of this ideology through the statements of the Business Roundtable, which is made up of CEOs speaking for 30 percent of all equity value in the United States. As late as 1981, the members of the Roundtable said that although shareholders must earn a fair return, "the legitimate concerns of other constituencies also must have appropriate attention." They believed that "the owners have an interest in balancing short-range and long-term profitability, in considering political and social viability of the enterprise over time."

By the mid-1990s, that sort of fair-minded balance was gone, replaced by fiduciary absolutism. "The principal objective of a business enterprise is to generate economic returns to its owners," the Roundtable now wrote. "The notion that the board must somehow balance the interests of stockholders against the interests of other stakeholders fundamentally misconstrues the role of directors."

Long-term sustainability be damned. Stakeholders be damned. Purpose be damned. Fiduciary absolutism has driven our corporations to focus on shareholders above all else. If sugary cereal serves shareholders better, Kellogg's has a sacred obligation to fill the bowl of every last child.

But even within the simplicity of this shareholder-focused framework, managers still have a problem: they have to serve shareholders, but which ones? Some shareholders want growth, others want dividends. Some seek risk, others seek stability. Some hold their stock for days, others for decades. Enter the efficient market hypothesis.

In its strongest form, the efficient market hypothesis says that a corporation's current stock price represents its true value. The stock price includes all relevant information and considerations, and therefore increasing the current stock price benefits all shareholders uniformly. Rather than worrying about balancing the interests of different time periods or risk tolerances, managers could now focus on a single number.

The corporate world is much simpler under fiduciary absolutism and the efficient market hypothesis. Fiduciary absolutism tells managers that they exist to serve shareholders. The efficient market hypothesis says that they do so by maximizing today's share price. This is the single objective function—the *one and only* thing that matters. Where else in life does this kind of unbalanced focus lead to better outcomes?

The result is the pernicious short-termism that plagues corporations today. In one recent survey, more than two-thirds of CEOs and CFOs of large public corporations said they faced pressure to maximize short-term returns at the expense of long-term growth. Consider the quarterly earnings that corporations report. The number of firms that provide forecasts for this short-term metric grew from 92 in 1994 to more than 1,200 by 2001. Two-thirds of CFOs said they had been pressured by other executives to misrepresent corporate results in order to meet earnings expectations. Four in five business leaders said they would decrease spending on research and development, advertising, maintenance, or hiring in order to hit those benchmarks. Over half of managers surveyed would skip carrying out a profitable project to do the same.

There's just one problem with the efficient market hypothesis: it's wrong.

It's what the economist John Quiggin calls a "zombie idea"—a concept that is intellectually dead but exercises broad influence nonetheless. The economist Luigi Zingales agrees. Whereas in the late 1980s, no other proposition in economics was as widely supported, today "it is hard to find any financial economist under forty with such a sanguine position." A company's stock price at any moment reflects only the price at which the last buyer and seller traded a share. That's it. There are things companies can do to increase the price today at the expense of long-term value or to benefit some shareholders more than others. Corporations have many different stakeholders even within their shareholder base and many different ways to serve them.

Yet the allure of a simple theory is strong. As the economist John Maynard Keynes once wrote, "Practical men, who believe themselves to be quite exempt from any intellectual influence, are usually the slaves of some defunct economist."

Here's the result of fiduciary absolutism: corporations see no purpose higher than profit, no duty greater than maximizing shareholder value. We have an economy that maximizes the one thing we shouldn't care about—short-term stock price—at the expense of everything we should. The real genius of fiduciary absolutism, however, is that it rests on a deeper moral justification. There is little in the principal-agent problem or the efficient market hypothesis to elicit much enthusiasm. So why has it become such a dominant way of doing business? Because, proponents argue, maximizing profit for shareholders is what benefits society most overall.

This is the invisible hand. "It is not from the benevolence of the butcher, the brewer, or the baker that we can expect our dinner," Adam Smith wrote in *The Wealth of Nations* in 1776, "but from their regard to their own interest." Our self-interest ends up benefiting others as if guided by an invisible hand. The modern rendition is that private vice makes for public virtue. As Keynes protégé E. F. Schumacher put it, we've come to think "the road to heaven is paved with bad intentions."

Maybe that's sometimes true. But as we'll soon see, the economy Adam Smith wrote about was drastically different from our own. A theory based on cottage industries and local owners works differently when applied to global corporations with institutional shareholders. Using the invisible hand to justify capitalism today is like using Jefferson's vision of a nation of small yeoman farmers to justify Monsanto and the Dole Food Company. We should be wary of sacred originalism in social science. Smith may have been right for his time, but, on this point at least, he is wrong for ours.

So maybe private vice makes for public virtue. Or maybe it makes for public vice. Maybe the invisible hand can cure all ills. Or maybe it causes them. As the prominent legal scholar Leo E. Strine,

Jr., former chief justice of the Delaware Supreme Court, has written, "If empowering short-term investors turns out to be optimal for our society and its human citizens, that seems like a very improbable and unsustainable triumph of the law of unintended consequences."

Let economists have their invisible hand. We'd rather have companies that feed our children a healthy breakfast, the way Dr. John Harvey Kellogg intended.

A BRIEF HISTORY OF CAPITALISM IN AMERICA

We've seen the ideological underpinnings of fiduciary absolutism, but why did they take root when they did? Why doesn't the invisible hand metaphor apply as well today as it did in 1776? For this, we have to take a quick sprint through the history of capitalism in America. We'll then see how closely Kellogg's followed this path and how a few corporations today are trying to create a different future.

There have been five eras of ownership in America, with the transitions between them forced by war, ideology, and innovation—both technical innovations such as the steam engine and financial innovations such as junk bonds.

1. **LOCAL OWNERSHIP (FOUNDING TO 1850s):** Most business is done on small farms and in cottage workshops.
2. **FOUNDER OWNERSHIP (1850s TO 1900s):** Industrial factories replace cottage workshops as the so-called robber barons build large corporate trusts.
3. **PROGRESSIVE ERA PUBLIC OWNERSHIP (1900s TO 1940s):** Trusts become broadly owned public companies that are beginning to be balanced by government regulation and unions.
4. **NEW DEAL ERA PUBLIC OWNERSHIP (1940s TO 1970s):** Managers balance the interests of all stakeholders; shareholders are mostly passive.

5. **INSTITUTIONAL OWNERSHIP (1970s TO TODAY):** Public companies are dominated by institutional shareholders such as mutual funds and hedge funds, which force a greater focus on short-term profits.

As we'll see, fiduciary absolutism came out of the transition from public ownership to the institutional ownership era.

LOCAL OWNERSHIP (FOUNDING TO 1840s)

When *The Wealth of Nations* was published, almost all economic activity occurred on small, locally owned farms and in cottage workshops. More than 90 percent of Americans lived in the countryside. In an 1812 letter to John Adams, Thomas Jefferson wrote, "every family in the country is a manufactory within itself, and is very generally able to make within itself all the stouter and midling stuffs for it's own cloathing & houshold use." As late as 1850, retailing was still completely dominated by locally owned stores.

At the turn of the nineteenth century, there were only 335 business corporations in all of the United States, most of which had been established for transportation projects such as canals and toll bridges. There were no middle managers in the economy as late as 1840. In fact, Adam Smith was worried that the few corporations that did exist in his time might put at risk the sort of economy that his theory was meant to describe. He believed that "the corporation undermined an ideal market economy composed of small, owner-managed companies," according to one modern scholar.

Finally, consumers knew what they were buying. There were no cars, no Cokes, no utilities—almost nothing that individuals couldn't make themselves. As late as 1894, the fabled Sears catalog still had only 322 pages of wares—more variety than most people were used to, yet quaint compared to Walmart's 150,000 items now on offer. Or Amazon's 500 million.

In the era of local ownership, an individual intuitively balanced his interest in making a profit with a local sense of accountability to his neighbors and a long-term interest in staying in his community's good graces. So it would remain through the eve of the Civil War. And then things changed. Fast.

FOUNDER OWNERSHIP (1850s TO 1900s)

US industrialization came on in a rush: railroads, electricity, telephones, cars, branded packaged goods, and department stores. All of them required large-scale production to be efficient. This, in turn, required capital to invest in heavy industrial products. The first corporations were those that needed to pool large amounts of capital for big up-front expenses such as train locomotives and factory machinery.

Toward the end of that era, large industrialists such as John D. Rockefeller, Andrew Carnegie, Andrew W. Mellon, and Cornelius Vanderbilt had put together massive conglomerates, consolidating most of the country's industrial base in only a few dozen companies. Though those corporations became massive, they were still run by their founders, called robber barons by some due to their monopolistic practices. Few companies in those days were public apart from the railways.

Those large corporations came to dominate everyday life—both as producers and as employers. By 1860, factory workers were replacing independent skilled craftsmen. The number of white-collar workers grew from less than 10 percent of the workforce in 1870 to nearly 20 percent in 1910. Markets became so integrated that by the outbreak of World War I, the price of a bushel of wheat was identical in Chicago and Liverpool. The era was characterized by laissez-faire economics, a hands-off policy of minimal government intervention. For the most part, government steered a course of benign neglect.

But the era's scale of production and distance between consumers and producers created room for abuse, as the meat industry shows. Before the Civil War, meat had been slaughtered on family farms and sold through the local butcher. But the industry was revolutionized by urbanization and refrigerated railcars. A local, small, and simple industry became the distant, massive, and complex meat industry of the Chicago stockyards, where millions of hogs and cows were slaughtered each year. This complexity led to abuses such as the packaging for sale of diseased and rotten meat documented in Sinclair's *The Jungle* and, after public outcry, to the Meat Inspection Act, the Pure Food and Drug Act, and the Bureau of Chemistry (later renamed the Food and Drug Administration).

As calls for regulation grew stronger and founders themselves sought to sell their massive corporations, we entered a new era.

PROGRESSIVE ERA PUBLIC OWNERSHIP (1900s TO 1940s)

President Theodore Roosevelt, who took office in 1901, thundered against "the malefactors of great wealth." He sought to contain the monopolistic practices that had emerged as capital had become concentrated in the hands of the few. Though Roosevelt broke up many of the trusts, big corporations remained. Increasingly, however, they were owned not by their founders but by widely distributed shareholders. The number of Americans who owned stock grew from 2 million in 1920 to 10 million in 1930 to 25 million in 1965.

The rising power of corporations was met by the rising power of government and labor. Four million workers staged 3,600 strikes in 1919 alone. By 1945, union membership had peaked at 35 percent of private sector employment. This had a direct payoff for union members through higher wages and better benefits. It also helped the rest of the American workforce, who saw their wages and benefits improve. Union membership has since declined to less than

7 percent today. The era also saw the birth of strong federal regulation, with the 1887 Interstate Commerce Act, the 1890 Sherman Antitrust Act, and the 1914 Federal Trade Commission Act among the earliest laws to be passed.

The spread of stock ownership in our economy is actually unique in the world. The United States and the United Kingdom are the only two countries to have their largest corporations held by a widely distributed group of shareholders. Whereas roughly two-thirds of public companies in Germany have a single shareholder who controls over half the company, in the United Kingdom it's only 2.4 percent. On the New York Stock Exchange, it's 1.7 percent.

The era was punctuated by the Great Depression, which marked the end of laissez-faire economics. Increasingly, citizens looked to government for leadership as referee, infrastructure builder, and distributor of the spoils of growth. By the 1940s, the Progressive, antimonopoly era of the first Roosevelt had become the New Deal, pro-stakeholder era of the second.

NEW DEAL ERA PUBLIC OWNERSHIP (1940s TO 1970s)

With the presidency of Franklin Delano Roosevelt, government assumed a larger role in protecting the people and public goods left behind by capitalism. The Food and Drug Administration led to Medicaid and Medicare, the Clean Air and Water Quality Acts, Social Security, and the other welfare programs and regulations of the New Deal Era.

Unions, still at their strongest, fought for benefits that had previously been the preserve of the rich. The number of workers covered by private pensions grew from 4 million in 1940 to 15 million in 1956, eventually covering 50 percent of workers by 1980. The same thing happened with what was then called hospital insurance, where the number of workers covered grew from 6 million in 1939 to 91 million in 1952. Those plans were often created under the premise that workers would stay at one firm until they retired—a model in which long-term reciprocal obligations between employers and employees, both formal and informal, were the norm.

The economy also soared, thanks in part to the Allies' victory in World War II. Not only did the war create a national sense of shared sacrifice and identity, it also left the United States the last industrial economy standing. After the war, the United States held 7 percent of global population but produced 42 percent of its manufactured goods, 57 percent of its steel, 62 percent of its oil, and 80 percent of its cars. The economy grew by nearly 4 percent annually from 1946 to 1973 (compared with 0 to 3 percent over the last ten years). Real household income increased by 74 percent over the same period. Economic prosperity was more widely shared than any time before or since.

As ownership continued to spread to millions of small, passive individual shareholders, corporate managers took control. Scholars call this the period of "managerialism," in which managers saw their role as balancing the varied and conflicting demands of all their stakeholders. Fiduciary absolutism had yet to take hold. A 1961 survey revealed that 83 percent of executives believed it was unethical to work only in the interest of shareholders.

The United Kingdom followed a similar transition over this period. "The position of shareholders," wrote one British economist in 1965, "which is sometimes presented by the ideologues of business in the image of a parliament telling ministers what to do, is in fact much closer to that of a highly disciplined army, which is permitted by law to riot against its generals if, but only if, rations should happen to run out."

This sense of balance in an economy of shared prosperity—shared at least among white males—marks an era that many capitalist reformers wish we could return to. But the era would soon give way as fiduciary absolutism began its reign.

INSTITUTIONAL OWNERSHIP (1970s TO TODAY)

The transition from managerialism to fiduciary absolutism today is the result of several factors: increased competition from abroad concurrent with a bear market in the 1970s that undermined Americans' confidence in business, the advent of hostile takeovers in the public markets in the 1980s, and, perhaps most important, a shift in ownership toward institutional shareholders.

Through World War II, the average US worker was five times as productive per hour as the average Japanese worker; by 1980, Japanese car workers were 17 percent more productive than American car workers. The same was true of steel: from 1956 to 1976, Japanese steelmakers went from 19 percent less productive to up to 17 percent more productive than their US counterparts. With free trade and global competition growing simultaneously, US corporations were forced to compete with those that had far lower standards with regard to workers and environmental regulation.

Meanwhile, in the 1970s, an Arab oil embargo quadrupled oil prices, triggering a period of stagflation in the United States: the economy stagnated while inflation ballooned to as high as 11 percent annually. From 1966 through 1982, the real return on the S&P 500 was 0 percent. Productivity growth slowed to roughly half the rate of previous decades. By 1971, the United States had an unfavorable balance of trade, the first time that had happened since 1893. Whereas in 1966, 55 percent of Americans had voiced "a great deal of confidence" in corporate leaders, by 1975, only 15 percent did so.

Around that time, the creation of high-risk "junk bonds" suddenly gave investment funds the ability to buy out entire companies on the public market, even against the wishes of their existing managers. In the 1980s, nearly a quarter of major corporations received a hostile or unwanted bid, and nearly three in five received a takeover offer of some kind. By 2000, half of the largest hundred industrial firms had either been taken over or gone bankrupt.

Finally, structural changes in how Americans saved for retirement led to a greater focus on short-term share price.

Until the 1970s, most pensions were structured as a defined benefit to be paid by the company. In this arrangement, workers' interests were in keeping the company strong, stable, and low risk so it could meet its pension commitments. Unfortunately, when companies went bankrupt, retirees could be left without the benefits they'd been promised. So the government set up new minimum funding rules, mandatory insurance contributions, and other regulations meant to protect pensioners. They protected retirees against some of the risks but also made those sorts of plans more expensive and risky for corporations.

But the government gave companies another option: Rather than promising retirees a pension in the future, a corporation can contribute to a tax-deferred retirement account today, such as a 401(k). Then it's up to the employee to save and invest that money. The employer is off the hook. If corporations wanted to convert from the old "defined-benefit" system to the new "defined-contribution" system, they could.

And most of them did. In 1981, 60 percent of pensioners relied on a defined-benefit plan. By 2001, that had flipped: 60 percent relied on a 401(k) or IRA. Total assets in 401(k)s grew from $700 billion in 1994 to $4.4 trillion in 2014. Again, this is unique in the developed world. In Germany and Japan, private pensions are still paid out of a corporation's cash flow. In many other developed countries, retirements are funded by the government. In France, retirees get 85 percent of their income from the government. In the United States, it's only 36 percent.

The United States' new reliance on defined-contribution retirement accounts had two major effects.

First, it meant that a comfortable retirement now relied on the returns of the stock market rather than the stability of a worker's employer. In 1977, 20 percent of households owned stock. Today, that figure is over 50 percent. Now everyone—union members and the rest—must worry about shareholder value to fund their golden years.

Second, it led to the rise of institutional investors, as employees turned to mutual funds to invest their retirement savings on their behalf. By 2000, there were nine thousand mutual funds in the United States, two-thirds of which had been launched in the preceding decade. Whereas institutions such as mutual funds controlled only 6 percent of equity in 1950, they controlled over 63 percent by 2016. These institutions compete for our business from quarter to quarter by trying to show a higher return than the competition's. Their interests end up being far more shareholder oriented and short term focused than those of the workers whose retirements they manage.

All the while, a divisive war in Vietnam and social upheaval at home exacerbated Americans' confusion and undermined their self-confidence. At that stage in US history, the ground for a reorientation of business priorities was fertile indeed.

It was in that context that Jensen and Meckling stated that the principal-agent relationship was the core problem of corporate governance, that the efficient market hypothesis gave managers a single number to maximize, and that the invisible hand gave everyone a justification to carry out their worst impulses. It was in that context that the seductive simplicity of fiduciary absolutism came to dominate our corporations. The culture and norms supporting the reciprocal obligations of the postwar period gave way at the same time that both regulations and union protection declined.

Real per capita income in the United States grew at an annual rate of 2.5 percent from 1950 to the early 1970s. From 1973 to 1995, growth fell to 1.8 percent per year, and it has since fallen further to 1.6 percent per year since then. Productivity growth has slowed. Business has also become less dynamic, with fewer companies being started and fewer corporations going public. Indeed, the number of public companies has fallen by half since 1996. Of course, correlation is not causation, and many of the factors that led to fiduciary absolutism—such as increased competition from abroad and stagflation—are themselves either causes or consequences of slower growth. Nevertheless, this is the record fiduciary absolutism must answer for.

It's a history we see reflected in many of the largest corporations that have lived through it, Kellogg's included. After being founder-owned for years, the company went public in 1952, selling shares to four thousand individual investors. Institutions now own 88 percent of the company. From founder ownership to public ownership to institutional ownership, all in step with the country at large.

Dr. John Harvey Kellogg made a product to serve his own patients each day. Today, shareholders aren't even permitted to tour the manufacturing plant. Investors in mutual funds might not even realize they own the stock.

DR. JEKYLL, MR. HYDE, AND TONY THE TIGER

Kellogg's began as a family affair. Though Dr. John Harvey Kellogg invented the cereal, it was his brother, Will, who created the empire.

From the start the two brothers saw the world in different ways. John was the nutritionist in chief, protective of his reputation as a doctor. "I have been interested in human service," he wrote. "Not in piling up money." Will, on the other hand, was the master marketer. Over time, he became more interested in capitalizing on John's medical background than in making his customers healthier.

As described in *The Kelloggs*, "Will pressed the doctor to expand the business . . . develop a national advertising campaign, sell cereal in grocery stores across the country and make some real money." For Will, it was always just a business. To grow their empire, Will innovated extensively around taste, adding substantial sugar and salt to the recipe for Corn Flakes against John's wishes. The rift left the brothers barely on speaking terms.

Together, they were the corporate equivalent of Dr. Jekyll and Mr. Hyde: the upstanding doctor transformed at night into the nefarious bandit. It was one company with a split personality—dueling motivations to serve others and to serve itself. Unfortunately for Dr. John, Mr. Will won. Kellogg's grew in his image.

After Will's death in 1951, his successor, Watson Vanderploeg, introduced a long list of sugar-laden cereals. He hired the advertising company behind the Jolly Green Giant and the Marlboro Man to invent characters such as Toucan Sam and Tony the Tiger. Kellogg's grew to be a $22 billion company, and it did so primarily by selling sugary cereal. Today, Kellogg's continues to sell $2.6 billion of breakfast food each year, much of that sugary cereals.

From a biological perspective, we're wired to eat what we can; overconsumption is a modern luxury humans did not evolve to resist. But we have a system that is complicit in exploiting this reality. Food companies are not trying to make us sick; they're just trying to sell us more and more product. "In this kind of investment economy," wrote NYU professor Marion Nestle, "weight gain is just collateral damage."

As one journalist put it, "Obesity is often described as simply a matter of managing one's calories and consequently cast as a lack of willpower on the part of an overweight individual. But it is probably more accurately understood in the context of a global food system that is incentivized by financial markets to produce low cost, high-calorie, unhealthy, and addictive foods."

It's not a question of personal lack of discipline; it's our food system, trapped under fiduciary absolutism's mandate to maximize profits at all costs. Kellogg's objective is creating profit for shareholders. If the by-product is childhood obesity, well, according to fiduciary absolutism, that isn't the company's problem to solve.

It is a strange but typical feature of our economy's split personality that although Kellogg's the company continues to generate so much ill health in children, the foundation Will Kellogg created simultaneously tries to undo it. The now $7 billion W. K. Kellogg Foundation was launched in 1930 to help disadvantaged children. It has even given a grant to none other than Curt Ellis—the same Curt Ellis who made *King Corn* and is now trying to get kids off sugary cereal.

So here we have it, the Jekyll and Hyde of our financial system: make money however you can, then use the spoils to remediate the very problems you have caused. It's "The Gospel of Wealth" all over again.

Fiduciary absolutism has created a world in which it seems normal for companies that are meant to nourish us to make us sick. This irony isn't limited to food. Think of all the pain that painkillers have caused this country. Or the financial stress of retirees caused by unscrupulous financial advisers. These are not morally ambiguous products—casinos, cigarettes, or porn. These are industries with a very clear underlying purpose. Yet because of fiduciary absolutism, they put profit above all else, often subverting whatever purpose they could have served.

Does that mean we've reached the end of history? Does that mean fiduciary absolutism will rule forevermore? Not if it's up to a pair of social entrepreneurs who are taking on the food supply system.

FROM GENERAL MILLS TO ONE MIGHTY MILL

When Jon Olinto and Tony Rosenfeld couldn't find something, they made it. Two entrepreneurs from Boston, they decided to launch a farm-to-table restaurant that could compete with fast food. Their restaurants, called B.Good, had chalkboards with the names and pictures of the farmers who supplied them. They served kale salads and quinoa, as well as comfort food in the form of grass-fed burgers and hand-cut fries.

For their first restaurants in greater Boston, they sourced zucchini for their seasonal salads from a fifth-generation family farm in New York, eggs from Nellie's Free Range in New Hampshire, cheese from a creamery in Vermont, and ice cream from down the road in Cambridge. Most of it bypassed the industrial food complex; they used local, unprocessed food whenever possible.

But there was one ingredient that, try as they might, they could not find locally: wheat. So in 2017, they launched One Mighty Mill with the subversively simple tagline "Wheat you can eat!"

Almost all the flour we eat in our bread and baked goods comes from one of only four companies: Archer Daniels Midland, Bunge, Cargill, or the Louis Dreyfus Company. Known collectively as ABCD, these companies control 90 percent of the world's grain trade. Their process for making flour is a case study in industrial food production.

For the last century and a half, flour was meant to be cheap, shelf stable, and lily white—originally thought to be a mark of purity. Few recognized that the industrial milling process stripped away the bran, the germ, and many of the essential nutrients that made grain healthy in the first place. One Mighty Mill was founded to make wheat a superfood again. Its first step was to partner with the sole US-based stone mill builder to build a 7,000-pound mighty mill. Next it opened a storefront on the main street of Lynn, Massachusetts, a down-and-out town on the outskirts of Boston. This is where they sell their bagels, pretzels, and whole flour–based foods, using flour from wheat harvested on small farms in Maine.

The business is tiny, but it's driven by a deep sense of accountability to consumers and communities—in some ways closer to the era of local ownership than to our own. One could easily imagine Dr. John Harvey Kellogg involved in an enterprise like this one. "You can't have healthy kids without healthy wheat," Olinto told us. "And because healthy wheat is perishable, it's hard to have healthy wheat without a local mill."

Consumers themselves have already begun pushing toward healthier breakfast options, and financial markets have started responding. Over the five years from 2014 through 2019, Kellogg's stock price has grown only 14 percent while the S&P 500 has increased by 76 percent. For all its harms, fiduciary absolutism is meant at least to maximize share price. As we've seen already in the examples in this book—and as we'll see in the empirical data presented in chapter 4—fiduciary absolutism often fails even to accomplish this much. When CEOs grow profit quarter after quarter, people say they have the Midas touch. We forget that the myth of King Midas was a tragedy. He died of starvation when his greed for gold made his world unlivable, a gilded shrine of misplaced values. In the long term, you reap what you sow.

Olinto and Rosenfeld are trying to use One Mighty Mill to find a way around the obstacles described above. What if, instead of profiting by making us sick, all food companies sought their profits through making us healthy? What if the economic value a corporation created was once again aligned with the social and environmental value it created? Olinto and Rosenfeld are trying to find out.

A FAILURE OF AMBITION

All doctors take the Hippocratic Oath. "I will remember that I remain a member of society," it reads, "with special obligations to all my fellow human beings." Many professions, from barbers to airline pilots, have their own codes of ethics.

There is no such code for the business world. Instead, business leaders read Sun Tzu's *The Art of War* and borrow phrases from the field of battle. For decades our food industrial complex has indeed made a killing. Ray Kroc, the legendary CEO of McDonald's, embodied a sort of ruthless approach to business. "This is rat eat rat, dog eat dog. I'll kill 'em, and I'm going to kill 'em before they kill me. You're talking about the American way—of survival of the fittest." Well, that's one view of the American way.

When Harvard Business School added a dedicated ethics module in the 1960s, one observer wrote that it sent the message "that ethics, like the caboose on a train, would lend a sweet symmetrical touch to business, but that the locomotive would haul just as well without it." Professor Rakesh Khurana and Dean Nitin Nohria have recently made efforts to further codify ethics at HBS, going so far as to propose a Hippocratic Oath for Managers. It included the pledge to "guard against decisions and behavior that advance my own narrow ambitions but harm the enterprise I manage and the societies it serves." Khurana hopes his students will learn that corporations are meant to improve society, not harm it. As the saying often attributed to President Theodore Roosevelt goes, "To educate a man in mind and not in morals is to educate a menace to society."

Fiduciary absolutism doesn't maximize profit; it maximizes the *profit motive.* And it doesn't maximize shareholder value; it maximizes today's share price. Fiduciary absolutism is its own ruler. It serves itself.

Worst of all, fiduciary absolutism often reduces meaningful employment to crass commercialism. Fiduciary absolutism tells everyone—explicitly—that they are just cogs in someone else's value extraction machine, that they are line items on an income statement meant to be minimized. This misunderstands and underestimates human motivation.

Capitalism's staunchest apologists call this dystopia our promised land. They hide behind the invisible hand and claim that this is the best of all possible worlds. It is a convenient view for those who seek to simplify our complex world into a single number. It is a convenient view for those who would rather excuse their behavior than change it.

Fiduciary absolutism reflects a failure of ambition and a *reductio ad absurdum.* The most important problem facing a food company is not the principal-agent problem. It never was. It's feeding people, nourishing them, and doing so in a way that will also generate prosperity for employees, managers, and shareholders.

We forget that corporations are social organizations originally designed to solve problems, not create them. They are capable of meeting the needs and reflecting the values of their societies while making money at the same time. Shouldn't we hold our corporations to a higher bar?

3

GOOD NEWS AND BAD NEWS

THE CAPITALIST DRIVER, ASLEEP AT THE SWITCH

John Louis, the chairman of the board of Gannett, leaned forward on the lectern as he looked up at the teleprompter at the back of the auditorium. "We now open the floor, hearing first from the auditorium." He paused briefly and smiled. For the third time, the scrolling text referenced the overflow room. "And there's no one in the overflow room." A few people laughed. Of course there wasn't. The auditorium itself was only half full.

Louis was presiding over the 2019 shareholders' meeting of Gannett, the United States' largest newspaper conglomerate. Each month, 126 million people read a newspaper Gannett owned: *USA Today*, the *Des Moines Register*, the *Cincinnati Enquirer*, the *Arizona Republic*, and nearly a hundred other local papers across thirty-four states. Gannett was a public company. Its ownership was divided into 114 million shares. Anyone who owned even a single share was invited to the annual meeting that May. But barely two hundred people showed up. As Louis opened the floor for comments, only one shareholder approached the microphone.

"I guess I get to kick it off again here," the seventy-eight-year-old man began. He held a copy of *USA Today* as he stood at a microphone in the aisle. "I'm from Detroit." John Lauve was an annual fixture at the meeting. He's a second-generation former engineer at General Motors.

"The issue I have is the only hope for this country is the news media, and you're not doing the job that needs to be done." He quoted Joseph Pulitzer, the eponym of the Pulitzer Prize: newspapers should focus on providing the news that the country requires, and then profit will follow. After Lauve spoke, two union representatives and an employee took their turns. Within ten minutes, it was over. The entire meeting didn't last an hour.

That was all a little odd, given that the meeting's outcome could decide the future of the newspaper industry in the United States. Alden Global Capital, a New York-based activist hedge fund, was trying to seize control of the company.

Alden wanted to combine Gannett with its own newspaper conglomerate, Digital First Media. As an activist hedge fund, Alden's investment strategy was to take an active role in the management of the corporations it invested in. By this point, Alden and Digital First were already notorious for what that management entailed. *Washington Post* media columnist Margaret Sullivan called Alden "one of the most ruthless of the corporate strip-miners seemingly intent on destroying local journalism." The prospect of Digital First's impending control over Gannett caused an uproar among readers, reporters, and politicians. They had already seen Digital First hollow out newsrooms in San Jose, Oakland, and beyond. In an open letter to Alden, Senator Sherrod Brown wrote, "Your newspaper-killing business model is bad for newspaper workers and retirees, bad for communities, bad for the public, and bad for democracy."

At the 2019 meeting, Gannett shareholders could question and debate and ultimately vote on who would lead their company.

Should they reelect the incumbent board of directors to stay the course or elect Alden's challengers for a new direction? Big question. But when Louis asked the audience who needed a paper ballot to cast a vote, only three hands went up. This is the state of capitalism in the United States: we have an economy in which the capitalist is king yet, even here at this critical moment, is nowhere to be seen.

Fiduciary absolutism supposedly ignores stakeholders to benefit shareholders. But shareholders are stakeholders, too. In all, 137 million Americans own stock, either directly or through an investment fund—that's 15 million more people than voted in the last national election. Yes, stock ownership, like all wealth in the United States, is distributed highly unequally: the wealthiest 1 percent hold 38 percent of stock market wealth, and the top 10 percent hold four-fifths. But what if we focused not on the wealthiest shareholders but instead on the typical shareholder? The median shareholder? We try to improve labor markets by benefiting the typical worker rather than the wealthiest; we should use a similar approach when we think about what it means to serve shareholders.

Of the 137 million Americans who own stock, the *typical* shareholder has a diversified set of stocks in a retirement account worth $65,000. The *typical* shareholder won't withdraw that money for decades. *Typical* shareholders make nearly all of their income from their work. They care more about their job than their stock return. They live in communities. They are citizens of this democracy. They would be best served by stakeholder-oriented corporations focused on long-term prosperity rather than short-term stock movement. Many critics of capitalism say we are living in an age of shareholder primacy, in which shareholders win while the rest of us lose. That's not true; all but the most short-term-oriented shareholders are ill served by fiduciary absolutism.

Most shareholders read the news, too, by the way. But with Gannett's future in the balance, would they wake up to defend it?

HIGH STAKES

A year before Gannett's 2019 meeting, in April 2018, the *Denver Post* published an editorial under the headline "As Vultures Circle, The Denver Post Must Be Saved." The paper had recently been forced by its owner to cut newsroom staff from more than 250 to fewer than 100, with another 30 cuts to come over the next two months. That owner? Alden Global Capital.

The editorial was a plea to Alden "to rethink its business strategy across all its newspaper holdings." The editorial board described the valuable role a newspaper such as the *Post* plays in a community, saying that it is a "necessary public institution vital to the very maintenance of our grand democratic experiment."

For many readers, that edition of the *Post* felt like the end of the line for the paper. In addition to the editorial, the issue carried opinion pieces with titles such as "The Stories That Might Not Get Told"; "Who Will 'Be There' When Journalists Are Gone?"; and "When a Hedge Fund Tries to Kill the Newspapers It Owns, Journalists Must Fight Back."

One fifteen-year veteran columnist wrote, "If we don't speak up now, then we will be destined to witness the demise of our city's largest and most essential news-gathering operations—and what would happen to democracy then? Who would then hold the powerful, Alden Global Capital included, accountable in the absence of a major metropolitan daily newspaper?"

Denver was just the latest victim. The *San Jose Mercury News* had suffered the same kind of newsroom cuts, as had the *St. Paul Pioneer Press* and the *Boston Herald.* That was the nightmare scenario that Gannett shareholders were supposedly at the annual meeting to discuss.

In December 2018, five months before Gannett's annual meeting, John Louis had met with a member of Digital First's board, Martin Wade III. The meeting was supposed to be about industry trends. After the meeting, as Wade stepped into the elevator, he mentioned the possibility of combining the two companies. Newspapers are a tough business; maybe joining forces would make it a little easier. Louis demurred. He said his focus was on finding a new CEO to replace Gannett's retiring incumbent. Wade agreed; that made sense. Perhaps they could meet again in 2019. The elevator doors closed.

Not a month later, the *Wall Street Journal* broke a story that must have shocked Louis: Digital First had quietly bought up 7.5 percent of Gannett's stock, and it intended to acquire the rest.

Gannett spurned the advance. Among other concerns, it cited Digital First's reputation. "Within the industry, [Digital First] has a well-documented history of significantly reducing editorial staff and cutting costs at its acquired properties," the company wrote in a press release. It addressed Digital First directly: "What commitments would you be willing to make to maintain newsroom staffing and ensure any acquired publications can continue to produce high-quality local journalism and contribute to the communities they serve?"

Digital First must have known that its reputation preceded it. In a letter to Gannett, it used the same phrasing about providing "a home for the Company's businesses and valued employees so they can continue to serve their local communities" twice on the opening page, not seven sentences apart. Four out of the five times the letter referenced employees, it referred to them as "valued employees." It was not until the second page that it referenced Digital First's industry-leading capabilities in "rationalizing labor costs," a business-speak euphemism for firing "valued employees."

In releases to the public, Digital First highlighted Gannett's underperformance. Gannett's stock had been in the doldrums for years. Though the S&P 500 had returned 36 percent since June 2015, Gannett's stock price had declined by 15 percent. Digital First thought that its way of running newspapers could return Gannett's stock to life.

NO MARGIN, NO MISSION

There's an old story of a nun, Sister Irene Kraus, who ran the $3 billion nonprofit Daughters of Charity National Health System with an iron fist, as obsessed with tracking costs and pricing services as she was with caring for people and fulfilling the hospital's public duties. She had a mantra: "No margin, no mission."

As hard as things were for the *Denver Post*, its local competitor had fared worse. The *Rocky Mountain News*, which had published daily from 1859 through 2009, had been shuttered. The paper had seen its circulation decline by more than half, from 426,000 in 2001 to 210,000 in 2008. At that point, its owner had put it up for sale. No one had been willing to buy it. With no buyer and $15 million of losses the previous year, the *Rocky Mountain News* had shut down. Two hundred journalists had lost their jobs.

Most newspapers are shrinking not because of the policies of their owners but because of paying customers—or rather the lack thereof. As technology has changed, fewer people are willing to pay for subscriptions and fewer advertisers are willing to pay for ads. Classifieds, once the cash cow of local papers, have been replaced by Craigslist and other sites. Newspapers' share of total advertising spending in the United States has been in decline for nearly a century as radio, television, and eventually the internet have provided advertisers more effective options.

Meanwhile, we, as consumers, are refusing to pay for them. Daily subscriptions nationwide fell by half, from 62 million in 1990 to 29 million in 2018. Just since 2015, 9 million people have dropped their subscriptions. Annual revenue from circulation and advertising in the newspaper industry has declined by $35 billion since 2005. That didn't happen in other industries. Say what you will about Digital First's ownership of the *Denver Post*, the newspaper is still standing. There is a fundamental gravity to economics. And right now, newspaper owners are trying to fly. The traditional business model is broken. *No margin, no mission.*

In 2016, Paul Huntsman, a scion of the billionaire industrialist Jon M. Huntsman, Sr., bought the *Salt Lake Tribune* from Digital First. Despite his best efforts to revive it as a profitable paper—despite paying employees out of his own pocket to keep it afloat—he announced in May 2019 that the *Tribune* would become a non-profit. Investors would be replaced by philanthropists.

Not all papers have suffered the same fate. In 2018, the *New York Times* earned $55 million in net income while employing 1,600 journalists, an all-time high. *The Guardian* earned a profit for the first time since 1998. It now makes a majority of its revenue online from readers, rather than in print from advertisers. It achieved that "minor miracle of net growth" in a surprising way: it didn't put up a paywall or require subscriptions, it just asked for money. At the end of every article, it asks readers for "as little as $1—and it only takes a minute." Over a single year, more than 600,000 readers gave money, more than half on a recurring basis. A subscription by any other name . . .

Meanwhile, while the twenty-four-hour news cycle has in many cases descended into partisan popular entertainment, the news service Reuters continues to hold down the fort of unbiased accuracy in its reporting. Undergirded by the Trust Principles, a kind of charter established more than eighty years ago that commits its reporters to "act at all times with integrity, independence, and freedom from bias," it, along with a handful of others, is a lonely outpost in a media world gone mad.

The future of news will not look like its past. The paperboy has gone the way of Paul Polman's milkman. But that doesn't mean that real journalism must disappear as well. Some start-ups have tried to revolutionize the industry by breaking out of its legacy structure. Vox, Vice, Axios, The Information, and others are rebuilding journalism for a new era, rather than adapting it from an older one. But at the moment, the state of the fourth estate is dire, and the political realm is suffering at least partly as a consequence of biased and unreliable news.

In some ways, news suffers from the tragedy of the commons. We all want news we can trust; we all want one—or two! or three!—vibrant local paper. We want investigative journalism and government watchdogs holding the powerful to account. We protest and picket when papers shut down or cut staff. But we don't subscribe to them. For those who watch these protests, it's hard not to think "If you care about news so much, then buy the damn paper."

Well, that's exactly what one local group in California did.

MEMORY LANE

Driving into the beachside hamlet of Half Moon Bay, an hour or so south of San Francisco, involves winding past nurseries nestled into the crook of green mountains and vineyards along the hills. As you come around the bend toward the ocean itself, the town starts to materialize, the vineyards are replaced with million-dollar bungalows, and the sunny, warm air is replaced with a cool fog rolling off the ocean.

Half Moon Bay is a picturesque town, a land lost in time. There's Cunha's Country Grocery & Second Floor Emporium on the main drag. There's City Hall on the corner, with the US and California flags waving over the sidewalk, a bench in honor of the district attorney outside. There are flyers for an Earth Day celebration. There's even a historic one-room jail, built in 1911 and now a venue for community events.

But the quaintest building of all sits directly behind City Hall: the offices of *Half Moon Bay Review,* the town's local paper since 1898. Though most other local papers along the coast have folded in the past few decades, *Half Moon Bay Review* lives on.

For decades, it had been controlled by a family-owned conglomerate of newspapers, Wick Communications. But in 2017, Wick decided to sell the paper and use the money to invest elsewhere in the company. It engaged a couple of newspaper brokers to canvass for buyers. Much of the initial interest came from small private equity firms, whose biggest questions usually revolved around personnel and salaries. Not a good sign.

Instead, a small group of local citizens decided to buy the paper themselves. The investment was led by Lenny Mendonca, a successful business consultant who had served on the board of *The Guardian* and would soon lead California's economic development under the newly elected governor. Among others, he was joined by Rich Klein, who had recently sold a tech company in San Francisco.

"I didn't want it to shut down. It was that simple," Klein said about buying the paper. He laughed. "You aren't going to make a lot of money at the end of the day." Clay Lambert, the paper's editor, agreed. "The main thing is knowing right from day one that we weren't in it for the money." They ended up restructuring the paper as a benefit corporation. The company's charter is explicit that the paper seeks to benefit the community in addition to earning a profit.

Making a profit is still important. "Okay, don't get me wrong, we're losing subscription revenues," Klein said. "Basing a newspaper on ad revenue doesn't work anymore." He was hopeful but clear-eyed about the fact that their work to save the *Review* hadn't ended with buying it from Wick. "You've got to make the economics work. The goal is to have a viable business, and that's . . . whatever." He thought for a moment. "Coming up with a business model that makes sense, that's ongoing yet."

The history of newspapers follows the history of fiduciary absolutism that we traced in the last chapter. In 1900, when the newspaper business was still flush with money, nearly all papers were under local, independent ownership. That was still true fifty years later for almost three in four papers. But today, after seventy years of deregulation and consolidation, only 15 percent of daily newspapers are locally owned. Instead, most are owned by a handful of large conglomerates—including Gannett.

It might surprise someone familiar with the newspaper industry to hear Gannett positioned as the righteous defender of quality news. Gannett itself bought sixty daily newspapers from 1945 through 1980, when newspapers were still very good businesses. Its flagship paper, *USA Today*, was derided when it was launched in 1982. Critics called it overly commercialized and said it lacked the depth of other national papers. Gannett has been accused of cost cutting and standardizing away unique local content. In some respects, Digital First is just the latest actor in a decades-long play. Through each stage, the connection among the company, its local community, and its mission has become more tenuous.

This is emblematic of what's happened throughout our economy. Ownership has become more distant, opaque, and intermediated. There was a time when most business owners were like the small-town shop, the local proprietor sponsoring the Little League team. The new owners of *Half Moon Bay Review* represent that sort of local, intimate ownership, as do Dixon at the Kyanite Mining Corporation and the founders of One Mighty Mill. Today, we all know that sort of ownership is rare not only in the newspaper industry but in every industry.

Unfortunately, distant ownership separates those who control corporations from the impact of their decisions. When local business owners fire employees, it is *their neighbors* they lay off. If they pollute the local river, it is *their river* that they are polluting. Externalities—the external costs such as pollution that companies can create—don't exist in the same way for local business owners. Distant ownership creates more room for amoral ownership. Maybe this is why 59 percent of Americans have a "very favorable" view of small businesses but only 17 percent do of large enterprises. Maybe this is why 75 percent of the general public trust shopkeepers but only 25 percent trust corporate executives.

The essayist, poet, farmer, and philosopher Wendell Berry has offered an alternative to fiduciary absolutism: the local economy. For him, it all starts with the local food movement, which eliminates the distant global supply chain that has separated us from the land and the farmers who work it. According to Berry, this distance has come much to our nutritional and ecological detriment. So with food and with all else, he advocates a movement to go local: buy local, work local, own local.

This would mean giving "everybody in the local community a direct, long-term interest in the prosperity, health, and beauty of their homeland," he wrote. This would mean that communities would not import products they could produce themselves or export products until local needs were met. By removing the distance between our economic decisions and their ramifications, we can eliminate much of what plagues our economy.

The problem, of course, is that you can grow more strawberries in the winter in Mexico than you can in New York. There are benefits to concentrating tech in Silicon Valley, apparel manufacturing in East Asia, or film in Los Angeles. For most industries, costs go down as scale goes up: pencils manufactured by the million are cheaper than those made by the dozen. Scale also allows for innovation. Companies with economies of scale can invest to decrease production costs or increase product quality. Going to a local economy would remove the benefits of both trade and scale, leaving our economy dramatically smaller and less productive—and as a result poorer.

That is not to say that Wendell Berry is wrong. What if the problem with capitalism isn't greed, corruption, or malfeasance? What if the root of capitalist dysfunction is the distance between owners and the companies they nominally control? *Half Moon Bay Review* is a small, locally owned company. It's meant to be profitable, yes, but only in pursuit of its deeper mission as a newspaper. In this, it's a far cry from the large, highly distributed corporations that are international in scope and impersonal in ownership. It's a far cry from corporations such as Digital First and activist hedge funds such as Alden Global Capital. Hell, it's a far cry from Gannett.

SHIFTING THE GAME BOARD

When Alden proposed taking over Gannett, it argued that it would mean a big payout for Gannett shareholders. It made its case on a website—SaveGannett.com—with a long presentation listing its grievances against Gannett's management.

Gannett issued its own press releases and a myth-busting memo to state its case. Louis, Gannett's chairman, called Alden's proposal "inadequate and not in the best interests of Gannett and its shareholders." In other words, Alden could go pound sand.

But as the *Wall Street Journal* had reported, Alden already owned 7.5 percent of Gannett's shares through Digital First. Gannett's board of directors had turned Alden down, but what if now, through Digital First, Alden could nominate and appoint a friendlier set of directors?

On February 7, three days after Gannett formally turned down the offer and three months before Gannett's annual meeting, the leaders of Digital First and Gannett met face-to-face. During the meeting, the Digital First managers tried to address each of Gannett's objections to the acquisition. At one point, the Gannett board and managers asked for a break. They needed to discuss the presentation and figure out what questions to follow up with. Within minutes, however, the game board shifted. During the break, Digital First's lawyers delivered notice that they, as shareholders, would be nominating six new directors to Gannett's board to replace three-quarters of those currently serving. The nominees included the current chairman of Digital First, who was leading the meeting.

Imagine this: You are representing a company in a negotiation. During a break in the negotiation, you discover that your counterpart is now attempting to replace you as the negotiator. The meeting was over. Gannett decided not to resume further in-person discussions. Two days later, Digital First sent Louis an email "expressing gratitude for arranging the meeting."

That launched the proxy fight that would be resolved at the Gannett annual meeting. It's called a proxy fight because most votes are voted remotely "by proxy," through a third party. Both Digital First and Gannett wanted shareholders to support their nominees for the board: the eight that Gannett had nominated against the six from Digital First. The eight directors with the highest votes would win. If Digital First was successful, it could effectively dictate the direction of the company.

It would be up to Gannett's shareholders to decide.

Here are a couple of questions for you to answer: Did any of the mutual funds or index funds you own hold shares in Gannett in 2019? If so, how was your vote cast?

ABSENTEE OWNERSHIP

We have a tendency in the United States to focus on the CEO when we think about a company. The CEO is not only the public figurehead but also the day-to-day decision maker. Gannett's annual meeting was held at its headquarters in McLean, Virginia. The building was filled with employees whose future would be determined by the decisions of distant shareholders. But as one employee sitting in the back row of the Gannett annual meeting told us, "I don't think many employees are even aware of what's happening." Rather than thinking about the shareholders or the board, most employees focus on the CEO.

Gannett's shareholder meeting revealed both the potential of corporate democracy and the sorry reality of what it has become.

All public companies are run like miniature republics, in which shareholders are the citizens and boards of directors their elected representatives. Rather than "one citizen, one vote," each share is accorded one vote. Those who own more shares get a greater say.

If you imagine that most shareholders carefully consider their votes each year, you wouldn't be wrong; you'd just be half a century too late. Remember, we are living in the era of institutional ownership. Whereas institutional investors—pension funds, endowments, mutual funds, and index funds—controlled only 6 percent of equity in the United States in 1950, they controlled 63 percent by 2016. Of our thousand largest corporations, institutions now control over 73 percent. And when it comes to shareholder votes, it's even more concentrated. Of all the shareholder votes in the S&P 500, 93 percent are cast by institutions. It's these institutions that set the direction of our economy today.

Who are these nameless, faceless institutional investors? They are pension funds for teachers and public employees; they are foundation and university endowments; and they are investment companies such as Vanguard and BlackRock.

In general, institutional investors take money from individuals and invest it on their behalf. For example, BlackRock manages $6 trillion, Vanguard manages $4.9 trillion, and State Street manages $2.5 trillion of money that's been entrusted to them by investors. These three institutions also control 90 percent of the index fund market. Together, they are the largest shareholder in almost 90 percent of the largest US corporations.

Though this money is managed on the investors' behalf, the institutionalization of stock ownership has concentrated power in those who lead these institutions. The CEOs of BlackRock, Vanguard, State Street, and their peers are increasingly assuming control as their firms grow. This is what Harvard Law School professor John C. Coates calls the Problem of Twelve. "It is not an exaggeration to say," he wrote, "that even if this mega-trend tapers off, the majority of the

1,000 largest U.S. companies will be controlled by a dozen or fewer people over the next ten to twenty years."

What's wrong with institutional investors dominating corporate ownership? The first problem is the way these institutions are compensated. Most institutional investors are focused almost entirely on outperforming their competition, and they are judged against their benchmark on a quarterly or monthly basis. They want the companies they own to hit their targets *today,* even if it might be bad for those companies in the long term. These institutional investors care more about the short-term performance of their portfolios than about building long-term corporate value.

The second problem is that institutional investors make ownership distant and anonymous. A retired public school teacher's money is managed by a multibillion-dollar pension fund, which allocates money to dozens of multibillion-dollar investment managers, each of whom then invests in dozens or hundreds of multibillion-dollar global corporations. That distant chain is now the most direct connection schoolteachers have to the corporations they're invested in.

As an activist hedge fund, Alden invests money on behalf of others—others who are hoping Alden will improve the value of the companies they're invested in. Two-thirds of all the capital behind hedge funds comes from institutional investors. At Alden, these institutions have included the California Public Employees' Retirement System, Coca-Cola's pension funds, a university endowment, and—strangest of all—the Knight Foundation. The Knight Foundation has the explicit mission to invest in journalism and "foster informed and engaged communities, which we believe are essential for a healthy democracy." Only our deeply intermediated financial system could afford such an irony.

The individual or pensioner sitting at the end of this chain of ownership never interacts directly with the companies he or she owns. This is why Gannett's shareholder meeting was so poorly attended. If individuals owned the stock of only a handful of companies, it might be possible—though perhaps still unlikely—for them to read their annual reports and vote on shareholder resolutions. But when you own dozens or hundreds of companies, it is impossible to keep track of them. Individual investors leave this up to their investment funds, which vote on their behalf.

Most large institutions employ a team specifically for voting and stewardship. At BlackRock, for example, thirty-three staff cover 14,000 portfolio companies. During the 2012 proxy season, BlackRock voted shares on almost 130,000 proposals at 15,000 shareholder meetings worldwide.

Enter proxy advisers, yet another layer of this mind-bendingly complex system. Many institutions lack the resources to do all this voting on their own. So they hire proxy advisers, firms whose sole purpose is to tell institutional investors how to vote. Two firms you've probably never heard of—ISS and Glass Lewis—control 97 percent of this market. Proxy adviser firms operate deep in the bowels of our corporate world and wield immense power. An ISS recommendation for or against a proposal can swing 10 to 30 percent of the vote.

This whole chain is meant to benefit owners—the ultimate savers at the end of this long chain—but it can fail in strange ways. Often, a single institution will end up voting both for and against a single resolution, such as voting for an acquisition through the target's shares but against the acquisition through the acquirer's shares. As former chief justice Strine said, of index funds repeatedly doing precisely that, "This is, of course, incoherent, stupid, and reflective of a lack of judgment being exercised by the index fund on behalf of its specific investors and their interests." It can get worse. State Street offers a gender empowerment investment fund (traded under the ticker symbol SHE) that didn't even vote to support several gender-based shareholder resolutions. We'll echo the good chief justice.

Let's review. Our corporations are run by CEOs, who are appointed by a board of directors, who are elected by shareholders. Nearly every vote cast in these elections is cast by institutions, many of which are just following the recommendation of proxy advisers. All of this makes the ultimate owners of these companies—retirees, pensioners, savers, and the like—totally voiceless. Whatever interest they might have in, say, saving the news industry fails to be reflected.

There are many reasons for individuals and organizations to buy and sell stocks. Daily trading allows our financial markets to operate effectively—providing liquidity for buyers and sellers, borrowers and lenders, corporations and investors all to trade at fair prices. The problem of absentee ownership arises when activists and long-term holders such as mutual and index funds don't rise to their governance responsibilities, outsourcing judgment and accountability to others.

It is this absentee ownership that enables fiduciary absolutism to dominate unchecked. In this global, remote, and deeply intermediated system, we have left only a single, tenuous connection between owners and their corporations: short-term profit.

MILGRAM-NIXON SYNDROME

Absentee ownership and fiduciary absolutism combine to create what we might call the Milgram-Nixon Syndrome of capitalism—an iteration of an idea that originated from one particularly colorful letter to the *Financial Times* in 2018. The Milgram-Nixon Syndrome helps explain why our corporations do not reflect our values.

In the 1960s, the psychologist Stanley Milgram performed a series of experiments that would become famous—and infamous—for showing how far people will go when they believe they're obeying orders. In his lab at Yale, Milgram would take two volunteers and designate one the "teacher" and the other the "learner." On one side of a wall, learners played a memory game while hooked up to an electric-shock machine. When they failed, teachers on the other side of the wall were directed by a scientist to deliver a painful shock. As learners continued to fail, teachers were told to deliver stronger and stronger shocks—even after they could hear that the learners were in extreme pain. Two-thirds of teachers continued to deliver shocks all the way up to 450 volts, marked "Danger: Severe Shock."

The whole thing was actually a ruse: the learners and the scientist were hired by Milgram. The shocks were fake and the events were scripted. But the teachers—the actual study participants—had no idea.

Milgram's study participants were just following orders. They could maintain that the responsibility for what they had done was not theirs but that of whoever had been giving the orders. Perhaps, in their view, they should be absolved because they had not ultimately been in charge. Milgram called this state of mind an "agentic state." Study participants could justify their behavior by believing that they were just acting as agents for someone else—someone else who was ultimately responsible for the actions done on their behalf.

This, if you'll allow the jump from the laboratory to the boardroom, is the excuse of managers who follow fiduciary absolutism. A CEO attempts to maximize shareholder value no matter the cost because that's what the shareholders demand. Jensen and Meckling argued that the principal-agent relationship was the core problem of corporate governance. Milgram turns this on its head: the problem isn't that agents don't follow orders; the problem is that they *do*.

The Nixon side of Milgram-Nixon Syndrome is even simpler: Richard Nixon's defense during the Watergate scandal was that he could not possibly be held to account for what his underlings had done. He was the president of the United States, and he should be absolved because he had delegated responsibility to others. This is the excuse of absent owners.

Milgram-Nixon Syndrome is the combination of the Milgram participants' "I was just doing my job" and Nixon's "How was I supposed to know?" Managers claim that they are just following orders to maximize profit. Owners claim that they are too distant to be responsible for what their companies do. The result is a financial system in which the buck stops nowhere. Responsibility is shifted and shifted until—*poof*—it disappears, leaving behind nothing but a mandate to maximize profit.

Small-business owners run their companies with their neighbors, employees, community, and environment in mind. Making a profit is essential, but it's contextualized within the business owner's broader goals. At *Half Moon Bay Review*, the connection between the owners and their company is direct, visceral, intuitive. It is their town that the paper covers. They are friends and neighbors of the people who work at the paper and read it each week. They'll have to answer for their decisions when they're shopping at Cunha's Country Grocery & Second Floor Emporium or sitting outside City Hall.

For most owners today, there is no such context. Wendell Berry's local economy could fix this, but at a tremendous cost. Maybe there's another way; maybe we need to reconsider who companies' ultimate owners really are.

SERVING THE TYPICAL SHAREHOLDER

Let us pose two more questions directly to you: If you have a 401(k) retirement savings plan, do you care about how it performs tomorrow or how it performs by the time you retire? And do you care about the performance of a single company in your portfolio or that of the entire account?

Fiduciary absolutism adopts the adversarial framework of shareholders against stakeholders. We imagine that shareholders care only about short-term profit. But as we've seen, over half of Americans own stock either directly or through an investment fund such as a pension or mutual fund. Let's look again at who exactly most shareholders are.

Typical shareholders are fifty-one years old, on average. Almost all of their financial savings are for retirement; 90 percent is held in tax-deferred accounts such as 401(k)s or IRAs. They can't even access those savings without a tax penalty for about a decade. Someone that age can expect to live another thirty years. So the typical shareholder today is investing for the long term.

They are also broadly diversified. For example, someone who invests in a Vanguard Target Retirement Fund automatically buys shares in more than 11,000 stocks and 15,000 bonds across every industry worldwide. Index funds like these own up to 30 percent of all public companies. Of shareholders who aren't invested in index funds, 90 percent are invested in mutual funds of some sort, usually putting two-thirds of their savings into such funds. Most mutual funds similarly invest in dozens of companies across industries and regions, owning a representative share of the global economy. So the typical shareholder today invests long term and is also broadly diversified.

If we pause here, we can already see that most shareholders could be considered "universal forever owners." They are universal in that their holdings broadly represent the universe of available companies. They are forever in that they are saving for so long that their interests are aligned with a nearly infinite time horizon. No man is an island, and neither is an investment portfolio.

But we can keep going. Though the typical shareholder has $65,000 worth of financial assets, his or her income and wealth depend mostly on his or her job. For most shareholders, 70 percent or more of their income comes from their current job. A system that serves these sorts of "worker owners" will be focused on building great companies over the long term. Former chief justice Strine wrote of

> the failure of our overall corporate governance system to represent faithfully the rational, long-term perspective of ordinary American investors who can only gain if public corporations make money the old-fashioned way, by imple-

menting sustainable strategies to sell products and services and not through edgy practices, accounting gimmickry, or never-ending cycles of spin-offs and mergers.

Typical shareholders care more about our economy's ability to provide good, high-paying jobs than its ability to provide high quarterly stock returns. They are also customers of companies and neighbors of companies. They live in the environment that companies may pollute. They are also citizens. When companies shirk paying their proper share of taxes, typical shareholders are left footing the bill. Indeed, of all the roles that typical shareholders play in society, that of shareholder is a relatively minor one.

And so typical shareholders care about the performance of the entire economy over the long term and its impact on stakeholders and the environment. If our corporations should serve shareholders, these are the shareholders they should serve.

Thinking of shareholders this way can have huge implications. Start with this: for the typical shareholder, many externalities disappear. If one company increases its profits by forcing its costs onto its suppliers, most shareholders do not benefit because they own both companies. They profit only when economic value is created through better products or more efficient processes—not financial engineering or zero-sum competitive practices. If another company increases its profits by polluting the environment, most shareholders are unhappy because it is *their* environment that's being polluted. If the government later cleans it up, it will be with *their* tax dollars.

There is a famous problem in economics known as "the prisoner's dilemma." Two bank robbers are caught and held in separate cells. If neither of them snitches, both will go free. If one snitches, he'll get three months in prison and the other will get a year. If both snitch, they will both get six months. The best strategy for both prisoners is to keep quiet and walk away. That's the cooperative strategy.

But each is worried that the other one will defect, and, because they cannot coordinate their strategy and don't trust each other, both end up snitching. That's the uncooperative strategy.

Here is the bizarre fact about capitalism in the United States today: if corporations were run for the long term, oriented around stakeholders, and driven by a sense of purpose, we would all—shareholders included—be better off. That's the cooperative strategy of the prisoner's dilemma.

But our corporations aren't run that way. Instead, many companies try to free ride off the efforts of others. Let everyone else worry about pollution or education or taxes, they figure; we'll just squeeze out a little extra profit this quarter. Every corporation run for short-term profit pushes us toward the uncooperative strategy, just like the prisoners snitching on each other.

What's so bizarre about this? We play both parts. Most of us are shareholders and stakeholders, investors in thousands of companies and citizens of the world they inhabit. We are trying to get a free ride and suffering the consequences. Our immense distance from and the multilayered intermediation of our financial system have masked this underlying fact. "Fishing with dynamite is a good strategy for an individual fisherman, for a while," writes Lynn Stout. "But in the long run, it is very bad for fish and for fishermen collectively." As a result, most shareholders are ill served.

The capitalist landscape of the United States today is populated by corporations that don't represent our ultimate interests as their owners, their employees and customers, and the citizens who oversee them. The problem is not one of values; it's the transmission of those values through our severely intermediated, supposedly amoral system.

If there's something broken about newspapers or something broken about capitalism, it must be because of us. That's the bad news. The good news? It's something we can fix.

THE LAST ACCOUNTABLE OWNER

John Lauve was the only shareholder to speak at Gannett's 2019 annual meeting. He's done so every year since he bought stock in the company. He flies in from Detroit, stays with family outside Washington, DC, and comes to tell management what he, as a shareholder, thinks should change. He believes that newspapers should focus on providing the news that the country requires, and then profit will follow. He speaks the way we'd expect a typical shareholder to speak.

Gannett's annual meeting was momentous, but it was not suspenseful. By the time the meeting began in May, the board already knew the results. Institutional investors had already cast their votes electronically. The room was only half full because the institutional investors who would determine Gannett's fate were also invested in thousands of other companies with thousands of their own annual meetings. It would be impossible for them to attend every meeting. And everyone else—the individual shareholders such as John Lauve or the concerned readers and citizens who could have been in the room for the price of a single share—didn't even know it was happening.

But in the end, Gannett's existing board prevailed. Though BlackRock supported Digital First—which meant that any absentee owner who invested with BlackRock unknowingly did so, too—Digital First didn't win a single seat.

After the meeting, the editor of the *Detroit Free Press* walked up to Lauve, empty coffee cup in hand. As editor, he was a powerful figure in Lauve's hometown. But here, he addressed Lauve as an equal. Their conversation was just one between a manager and an owner, talking about how to deliver on the corporation's mission while still earning a profit. It echoed the conversations happening across the country at *Half Moon Bay Review*.

We've seen how distance and intermediation in our financial

system have bred absentee owners that allow fiduciary absolutism to run rampant. These absentee owners have become the Nixon of the Milgram-Nixon Syndrome. But we must now recognize that most shareholders would be better served by corporations run for long-term stakeholder benefit. The question is, how can we reform our economy so that this will happen? That's the subject of the rest of the book.

John Lauve is an uncommon shareholder, exercising the right he has to attend the annual meeting and tell the board he elects and the managers it hires exactly what he wants to be done. Amazingly, he owns only two shares of Gannett, worth about fifteen dollars. His cab ride to the meeting probably cost more than his ownership stake in the company.

Why doesn't Lauve own more than two shares? "It's not something you put your money into." He smiled, locking eyes through his bifocals. "Poor return. There's no future in it." Does he at least subscribe to his local paper? "I got so exasperated," he said, "I canceled."

4

FIGHTING FOR CORPORATE SOCIAL RESPONSIBILITY

FROM SUPERFICIAL WINDOW DRESSING TO REAL ACCOUNTABILITY

Even now, more than a decade after the Lehman Brothers collapse dragged the world into financial crisis, one of its architects still believes the work he was doing was good. Not neutral. Not morally permissible but straddling the line. Good.

"We were providing capital to people who had been cut out of the financial system so they could buy houses. This was a problem of risk management, not malicious intent," he insisted.

"But it's true, our culture was to fight hard on the financial battlefield. To do what was necessary to win and then to distribute the spoils of battle in the form of charity afterward. We always thought that winning was the most important thing." That desire to win led him to help develop the high-risk financial products that proved toxic to a society already gorged on debt.

Chuck Prince, the CEO of Lehman's rival Citigroup, put it this way before the financial crisis: "When the music stops, in terms of liquidity, things will be complicated. But as long as the music is playing, you've got to get up and dance. We're still dancing."

Before its collapse, Lehman Brothers had a remarkably forward-thinking corporate social responsibility program. CSR denotes a broad, voluntary set of activities including corporate philanthropy, organized volunteering, environmental efforts, and diversity programs. Today, some call it "sustainability," while others refer to it as "ESG," denoting its various environmental, social, and governance metrics and programs. By whatever appellation, it connotes a kind of self-regulation to benefit society and improve a company's reputation.

Lehman's 2007 annual report told shareholders, "Strong corporate citizenship is a key element of our culture. We actively leverage our intellectual capital, network of global relationships, and financial strength to help address today's critical social issues." It also reported that the company had hired Theodore Roosevelt IV, a great-grandson of the trust-busting, conservationist president, as the chairman of its Council on Climate Change.

Lehman president Joe Gregory was known for his emphasis on diversity and inclusion. As one contemporary recalled, "Great rallies were staged . . . with free cocktails and hors d'oeuvres served for up to six hundred people, all listening to Joe or one of his henchmen pontificating. 'Inclusion! That must be our aim!' he would yell." Rumor has it that the director overseeing diversity initiatives earned more than $2 million per year—and that the diversity division had a bigger budget than the risk management division did.

Yet despite those well-intentioned efforts, Lehman is widely blamed for bringing down the global financial system. How do you square a culture of corporate social responsibility with precipitating the Great Recession—the millions of families displaced and jobs lost?

In the same 2007 annual report to shareholders, Lehman posed a question. At the head of a breathless section listing the various charities to which the company had donated, management wrote: "Where will you make your mark? At Lehman Brothers, that is what we ask all of our employees."

Today, many people look to CSR as a panacea for our ailing society. Corporations have been happy to play their part, publicizing every dollar they spend on CSR in press releases, glossy reports, and expensive advertising. Eighty-six percent of S&P 500 companies issue sustainability reports of some type, up from only 20 percent in 2011. They talk about the environment and the importance of sustainability. They talk about focusing on all of their stakeholders—employees, customers, and communities. They talk about corporate citizenship and shared prosperity. They talk.

But the average company spends less than one-tenth of 1 percent of its revenue on CSR. The companies on the Fortune Global 500 had combined annual revenues of $32.7 trillion in 2018. For comparison, the entire US gross domestic product that year was $20.6 trillion. These companies dominate our world, but not through their CSR departments.

CSR is often small and superficial, a Potemkin village constructed to appease capitalism's critics. It's often voluntary and self-reported, lacking any real accountability. It often feels more like a corporation buying an indulgence from society than actually reforming its business to do good. In 2018, Chevron announced it would invest $100 million in lowering emissions through its new Future Energy Fund. That same year, it spent $20 billion on capital and exploratory investments in oil and gas. It's hard to argue that you're committed to change if you're spending 99.5 percent of your budget doing the same old thing.

Other corporations, however, go deeper. They build their business models to serve employees, customers, communities, and the environment as well as shareholders. They embody a set of values and look for profitable ways to serve all their stakeholders. For them, CSR isn't a marketing campaign; it's how they run their companies. Capitalism would look far different if this were what most companies looked like. Unfortunately, they are the exception, not the rule.

In this chapter, we'll see examples of both the superficial and the fundamental kinds of CSR at some of our largest corporations: Goldman Sachs and Philips, Coca-Cola and CVS, Wells Fargo and the French insurance giant AXA. We'll see pioneers pushing to make CSR core to business strategy, standardizing the way it is measured and requiring corporations to report on it. A century ago, public company financial statements underwent a process to become transparent, standard, mandatory, and audited. This created a sense of accountability that corporate social responsibility desperately needs today.

As for Lehman: after it had gone bankrupt, it received one last accolade for its responsibility—an award in 2008 for a ten-year mentoring project at an East End secondary school in London. The firm had been reduced to ashes and a handful of ongoing lawsuits, a symbol of Wall Street's excessive risk taking and greed. Nevertheless, here it was, receiving posthumous praise for its outstanding corporate social responsibility.

GOLDMAN SACHS AND THE CRUSADE FOR GENDER EQUALITY

Around the time of the Great Recession, Goldman Sachs had something of an image problem. In 2010, Matt Taibbi wrote an article for *Rolling Stone* that included the following colorful indictment:

> The first thing you need to know about Goldman Sachs is that it's everywhere. The world's most powerful investment bank is a great vampire squid wrapped around the face of humanity, relentlessly jamming its blood funnel into anything that smells like money.

Like Lehman, Goldman was in deep on subprime mortgages. But unlike Lehman, Goldman was properly hedged—by betting *against* the very mortgages it was underwriting. "That's how audacious these assholes are," Taibbi quoted one hedge fund manager as saying. "At least with other banks, you could say that they were just dumb. . . . Goldman knew what it was doing."

With vampire squid–level brand issues, Goldman was desperately in need of something positive on the PR front. It launched the 10,000 Women initiative, designed to educate women entrepreneurs in faraway places and provide them with access to capital. It also gave Goldman some virtue to trumpet.

In partnership with the World Bank, the program has committed $1 billion to 50,000 women in twenty-six emerging markets to date. These are women such as Ayodeji, who was able to start a catering business in Lagos, Nigeria. Goldman makes much of these successes: "As a financial services company, we know that gender equality isn't just important, it's an economic imperative."

If Goldman believes that gender equality is an economic imperative, how has it changed its core business? In 2018, only two women sat on its eleven-person board of directors. This put Goldman 358th out of the Fortune 500 on board gender diversity. Only four women sit on its thirty-person management committee. In the United Kingdom, where firms must report their gender pay gaps, Goldman reports that its female employees earn 55 percent less than its male employees on average. Goldman argues that for the same job, women make 99 percent of what men do across the firm. The 55 percent gap only reflects the fact that there are more men than women in senior positions. Well, exactly.

The 10,000 Women initiative seems like a wonderful use of money addressing a very real need. An investment of $1 billion over the last five years is a real expenditure. But in 2018 alone, Goldman earned nearly $37 billion in revenue off $933 billion of assets. It spent $12 billion on salaries and benefits for its employees—those same employees who experience a 55 percent pay gap by gender.

To be fair, Goldman managers insist that gender and diversity are a top concern. According to one former manager, when a woman and a man compete for a job, the firm always goes with the woman if the candidates are close. And the company's struggle for diversity is shared across the field of finance. Goldman announced in 2019 that it is targeting 50 percent women for all entry-level hires and is focused on improving the representation of women in senior positions as well, though it's got a long way to go.

"Purpose & Progress," Goldman's environmental, social, and governance report for 2017, highlighted the company's sustainable growth, results for all stakeholders, and being an "influential, positive force in the world." It talked about gender pay equity and highlighted the 10,000 Women initiative as evidence of the company's commitment.

Compare this to that same year's annual report to shareholders, in which the terms "gender," "women," and "female" come up in only two contexts: first, in a footnote stating that "words denoting any gender include all genders"; second, in reference to a class action lawsuit filed by three female former employees alleging that Goldman "systematically discriminated against female employees in respect of compensation, promotion and performance evaluations."

Like many corporations, Goldman has developed a split personality. It's what the late William T. Allen, a former chancellor of the Delaware Court of Chancery, which handles the internal affairs of companies, called "our schizophrenic conception of the business corporation." We alternately understand corporations in two contradictory ways. First, we see them as the private property of shareholders, their only purpose to maximize their financial value. Second, we see them as social entities "with purposes, duties, and loyalties of their own; purposes that might diverge in some respect from shareholder wealth maximization." Corporate law itself oscillates between these two conceptions. This is exactly the Two-Buffett Paradox we saw in chapter 1. Goldman is trying to respond to two contradictory demands.

And so Goldman issues two annual reports, one about its business for shareholders and the other about its social and environmental impact for the rest of us. Both are signed by David Solomon, its CEO. Both offer a picture of what Goldman does, why it exists, and how it affects the world. But there is little connection between the two.

CSR can be a vehicle for clarifying the core purpose of a corporation. It can also be window dressing, an afterthought meant to appease the activists. Our largest corporations have already adopted the language and posture of CSR. In addition to the 86 percent of the S&P 500 issuing sustainability reports, half of the companies on that list referenced environmental, social, or governance topics during their quarterly earnings calls. Nine in ten public company leaders now say that serving stakeholders beyond shareholders is important to them.

Yet capitalism remains unchanged. Of those company leaders who believe that serving stakeholders is important, 96 percent say they are satisfied with the job their company is already doing to serve them. Back in 2005, the deputy editor of *The Economist* wrote, "The movement for corporate social responsibility has won the battle of ideas." Maybe so. But it's losing the war of substantive action. CSR risks becoming no more substantive than the typical episode of *Undercover Boss*, at the end of which CEOs appear saintly in their magnanimity by paying an employee's tuition or medical bills but leave their companies otherwise unchanged.

THE STATE OF CSR TODAY

If you watched the 2018 Super Bowl, you saw a tear-jerking commercial by Hyundai.

Unwitting fans who happened to be Hyundai owners were dramatically removed from the stadium entrance and taken to a small, windowless room. There they watched a video of Hyundai helping child cancer survivors. "Every time you buy a Hyundai, a portion of those proceeds go to childhood cancer research," a narrator stated. One mother, looking into the camera, told the fans, "You helped save my child's life and the lives of so many children." A door opened, and then the big reveal came: the fans were introduced to the survivors in person. There were hugs. There were tears. During one embrace, a survivor said into a fan's ear, "I just want to thank you for owning a Hyundai."

Let's not denigrate this project. It's noble to fight childhood cancer, and this funding is money well spent. But let's be clear: Hyundai is a car company. It sells cars. It's wonderful that it also donates money to charity, but cancer funding is a long way from its core business.

During that same Super Bowl, you may also have seen Budweiser's attempt to cast itself as a disaster relief organization: hydrating the afflicted by providing canned water instead of beer. NBC charged $10 million for a minute-long commercial that year, a new record. Rather than using their precious time to advertise their products, many companies used it to show how much they cared.

The problem with CSR today, as it's practiced by the vast majority of corporations, is that it is unrelated and marginal to the corporation's core business. It's not that these programs don't do good. It's just that whatever good they do is small compared to the principal impact the corporations have. Kellogg's and McDonald's, two companies at the center of our food system, have both been featured on *Ethisphere Magazine*'s "World's Most Ethical Companies" list. We don't doubt that they were deserving based on the metrics the Ethisphere Institute tracks. But we do question whether the metrics put enough emphasis on the health impact of eating Froot Loops every morning and a Big Mac every afternoon.

Take Coca-Cola. In its advertising, it has highlighted its donations to help the children of battered women and its partnership with the Special Olympics. These are wonderful projects, though admittedly far from the core business. A little closer to home, it became the first Fortune 500 company to reach water neutrality, returning 257 billion liters of water to nature and reducing total water use in manufacturing by 16 percent. Improving the environmental impact of production starts to feel like more fundamental change. Nonetheless, the company still makes most of its money by selling carbonated sugar water. The average American drinks about 150 liters of the stuff every year. Corporate philanthropy and even more sustainable production won't change Coca-Cola's core product and what it does to people's health.

We used to call type 2 diabetes "adult-onset" diabetes because, as recently as 2000, it was almost unheard of in children. That is sadly no longer true; if you're a black child in America, the chance that you'll get type 2 diabetes has increased by well over 50 percent since then. Meanwhile, researchers have found that companies including Coca-Cola specifically targeted "Black and Hispanic consumers with marketing for their least nutritious products, primarily fast-food, candy, sugary drinks, and snacks." Coca-Cola executives went so far as to channel funds through industry organizations to researchers who would discredit the idea that sugar in beverages is a contributor to obesity, instead emphasizing the importance of physical activity in controlling weight.

We shouldn't let Coca-Cola's support for the Special Olympics—or even its admirable sustainability programs—disguise the fact that its core products are consumed in a way that's making its customers sick. Under Muhtar Kent's leadership from 2008 to 2017, the company also grew its sales of water and other better-for-you drinks to nearly a third of revenue as a matter of corporate strategy. This has probably had more impact on society than the rest of its CSR efforts combined.

Managers reflect their priorities in how they organize their companies. CSR is rarely a responsibility of the corporate strategy team, the operations team, or even the risk management team. Instead, it's relegated to the marketing or investor relations department. Companies seem to care about corporate social responsibility only insofar as it affects their perception by the public. But capitalism doesn't have a public relations problem. It has a purpose problem.

MOVING CSR TOWARD THE CORE BUSINESS

In the nineteenth century, when William Lever launched what would later become Unilever, he offered a promotion: he would donate £2,000 to the charity that received the most votes from customers. But the promotion failed to garner much enthusiasm; the customers didn't vote. So he changed course. Rather than donating a portion of the company's profits, he worked on changing the core of its operations: the conditions of its workers, how it bought raw materials, and the products it sold.

"Core" as a business term has become prevalent over the last twenty years, due in large part to business consultant Chris Zook's book *Profit from the Core: A Return to Growth in Turbulent Times.* First published in 2001, the book has a simple thesis: successful companies develop a strong, well-defined core business that drives their growth. It's a useful heuristic in the context of corporate social responsibility: if the change is at the core of the business, it's powerful. If it isn't, it's window dressing.

This notion has some high-profile advocates, including Larry Fink, the chairman and CEO of BlackRock. With $6 trillion of assets under management, when Fink talks, the business world listens. In 2018, he changed the conversation around CSR.

Each year, Fink writes a public letter to CEOs. For many years, the letters garnered little attention beyond the business press. In 2012, he discussed corporate governance; in 2014, innovation. But in 2018, he wrote a longer letter under the title "A Sense of Purpose." Within days, it was being reported in the *New York Times*, debated endlessly on Wall Street, and assigned in business school classes.

In his letter, Fink expressed hope that the private sector would make up for the inability of government to solve our social problems. But his proposal went beyond hope: he argued that society *demanded* that companies serve a social purpose that would benefit all their stakeholders. Fink's letter did not advocate giving more money to charity or supporting volunteer programs. Instead, it focused on the inextricable link between long-term growth and sustainability:

> Without a sense of purpose, no company, either public or private, can achieve its full potential. It will ultimately lose the license to operate from key stakeholders. It will succumb to short-term pressures to distribute earnings, and, in the process, sacrifice investments in employee development, innovation, and capital expenditures that are necessary for long-term growth.

We can see what more fundamental CSR looks like through three very different companies: CVS, AXA, and Wells Fargo. Each reveals the power of making CSR core to the company's strategy rather than relegating it to a marketing afterthought.

Our first example is the way CVS changed its approach to selling tobacco. In February 2014, its CEO, Larry Merlo, released a video announcing that over the following year, CVS would stop selling cigarettes and tobacco products in all of its 7,800 stores. In his message, he positioned CVS as a health care company, saying that it wanted to play an expanding role in improving its customers' health while reducing the cost of health care in the United States. "Cigarettes and tobacco products have no place in a setting where health care is delivered," he said.

At the time, CVS was selling $2 billion of tobacco products each year. Dollar stores had recently introduced tobacco products, drawing traffic away from other retailers. When CVS made the change, its merchandise sales fell by 8 percent.

But over time, the company helped to reduce smoking. Smokers didn't start buying cigarettes elsewhere; they bought 95 million fewer packs overall. Total cigarette sales declined by 1 percent in the states where CVS had a significant presence. CVS didn't just donate some portion of its profits to a cancer foundation or promote smoking cessation in its stores; it went to the core of its business model—the very product categories it sells—and made a change. Lives were changed as a result.

CVS was not the only company to reconsider tobacco's role in its core business. AXA went a step further. AXA is one of the largest insurance conglomerates in the world. Headquartered in France, it has $992 billion in assets, 171,000 employees, and 105 million clients across 61 countries. In 2016, it recognized a contradiction in its business. "As a health insurer, we see every day the impact of smoking on people's health and wellbeing," said CEO Thomas Buberl. "Insurers should always be part of the solution rather than the problem when it comes to health risk prevention."

Céline Soubranne, the head of AXA's corporate social responsibility department in France, described the contradiction: "It was nonsense. On the one side, you earned money thanks to the tobacco industry. On the other side, you lose money because of the tobacco industry." AXA decided to resolve the contradiction. "As health is one of our key business priorities," said Buberl, "we adopted the bold decision to become the first insurer to stop providing insurance to tobacco manufacturers." The company went on to stop providing capital to tobacco companies, divesting €2 billion in loans and equity investments from its portfolio.

As an insurance company, AXA specializes in risk. It is natural for the company to reorient its business around solutions that minimize global risk, whether that involves ostracizing tobacco or fighting climate change. Though AXA has a separate CSR department, even Soubranne, the head of that department, hopes to see it integrated into the rest of the company in the future. "It's my dream that we will be fully integrated. I don't think my function will be necessary in the future."

We began this chapter by looking at financial companies such as Lehman and Goldman; now let's apply this thinking to another behemoth of our financial system: Wells Fargo. We'll set aside its larger scandals and focus on one recent press release, when Wells Fargo announced three ways it was serving people with disabilities. First, it had donated $100 million to fifteen thousand nonprofits that help those with disabilities. Second, it had made an effort to hire people with disabilities, employing eight thousand people who self-identify as having a disability. And third, it had reoriented some of its banking services to serve those with disabilities, providing hands-on financial education and help with buying a home or starting a business. Those were all good, worthy efforts, to be sure, but we can now rank Wells Fargo's efforts according to how close they get to reforming its core business of banking. Philanthropy is good. Hiring different people is better. But changing its core products to better serve its customers is best of all.

The difference between authentic CSR that is core to a company's strategy and almost-authentic CSR that is marginal is the difference between hope and cynicism, between a better theory of the firm and an empty marketing exercise that exacerbates the underlying distrust in our companies. This leads to a counterintuitive conclusion: if a company really cared about corporate social responsibility, it would fire its CSR team and focus on strategy instead.

MOVING CSR TOWARD STANDARDIZED METRICS

On the current frontier of corporate social responsibility, it can be hard to tell the pioneers from the charlatans, the cynics, and the fools. Even when a corporation has brought CSR into the core of its business, it can be hard to assess how well it is doing or how it compares to its competitors. ESG metrics—the way companies measure the impact of their environmental, social, and governance programs—are notoriously squishy. ESG reports overflow with precise measures, but most of them are self-reported, none of them is standardized, and almost all of them are impenetrable to nonexperts.

Look up the five largest companies in the world by revenue, and every list will be the same. Look up the most responsible, and there's no agreement. In 2018, only one company made it into the top five of both *Barron's* 100 Most Sustainable Companies and *Newsweek*'s Top 10 Green Companies. If we look at a company's credit rating, there is an almost perfect correlation between how different ratings agencies evaluate them. But between a company's various ESG ratings, the correlation is about halfway between perfect and zero.

In 2018, the *Wall Street Journal* published an article titled "Is Tesla or Exxon More Sustainable? It Depends Whom You Ask." FTSE Russell, MSCI, and Sustainalytics—ESG ratings agencies commonly used by investors—gave divergent assessments. The author offered the natural conclusion: "Investors should not treat ESG scores as settled facts."

No one would say that revenue or cash flow is in the eye of the beholder, but today, many would say that of ESG ratings. Although four in five CEOs believe in the importance of demonstrating a commitment to society, none of them has a reliable set of metrics to guide this commitment—leaving both executives and investors in the dark. In a world where fiduciary absolutism leaves room for gaming just about *anything*, ESG metrics have left the door wide open to abuse.

As a 2019 study showed, the reasons ratings differ are legion: different criteria (should we include lobbying?), different measures (should we assess labor according to employee turnover or according to the number of labor-related lawsuits?), different weightings (is water use more important than the minimum wage?), and—underlying all this—inconsistency in how the raw data are collected, consolidated, and compared. Most ratings rely heavily on company disclosures, often found in annual ESG or CSR reports. Since company disclosures are entirely voluntary, self-reporting bias is unsurprising. To counteract reporting bias, ratings agencies such as Truvalue Labs and RepRisk source "objective" external data to create their assessments, using artificial intelligence to comb through news articles and social media.

Compare all this with financial accounting. In the United States, all public companies comply with Generally Accepted Accounting Principles (GAAP), which are set by the Financial Accounting Standards Board (FASB) as overseen by the Securities and Exchange Commission (SEC) and audited by private accounting firms such as EY and PwC. It's an alphabet soup of accountability, but for the most part, it works. Though each company is unique, all financial statements are reported according to the same standards.

In 2011, the Sustainability Accounting Standards Board (SASB) began working to provide a new, standardized ESG reporting framework to mirror the success of GAAP accounting. One risk with any reporting framework is that it will be too onerous, focused on too many irrelevant factors in too much excruciating detail. Against that, SASB has approached ESG reporting through the lens of "materiality."

SASB wants companies to focus on disclosing ESG metrics only in areas that are material—what we've been calling core—to their businesses. For example, data security and customer privacy are core issues for tech companies but not infrastructure companies. Wastewater management is core to mining companies but irrelevant to health care companies. For each industry and subindustry, SASB focuses companies on the ESG metrics that matter for them. These standards were developed in collaboration with investors and accountants to create a framework that is practical and useful.

There is a risk that in an attempt to please everyone, however, SASB reporting has become too complicated to be intuitive and is consequently open to gaming. We know what our largest social issues are, even if we disagree on their relative importance. We'd prefer to see a small number of common metrics reported on by all companies in the same way—for example, satisfaction scores for employees and customers and overall carbon emissions. But SASB's efforts at standardization are a move in the right direction; it is a platform that can be built upon and improved over time.

As these metrics become more central to an organization's business—more directly linked to its long-term risk and performance—they begin to shed their do-gooder history. Terms such as responsibility are replaced by sustainability, acronyms such as CSR are replaced by ESG. The terms change, but the concept is the same. The king is dead. Long live the king!

MOVING CSR TOWARD MANDATORY REPORTING

With SASB and other bodies, such as the nonprofit Global Reporting Initiative, pushing for standardization, there remains one other significant gap between financial reporting and ESG reporting. Whereas standard, audited financial reporting is required for all public companies, ESG reporting remains entirely voluntary.

Our goal should be to make corporate social responsibility core to a company's purpose—to everything it does. To achieve this goal requires simplified, standardized, audited, and mandatory reporting.

According to a former head of the Securities and Exchange Commission, Mary Schapiro, two-thirds of all "comment letters," or requests for clarification from companies, are now about ESG disclosure. "I think there's increasing appreciation by the agency that these are important issues," Schapiro said, "and that they have greater potential to have financial impact on companies as time goes on." If shareholders keep pushing for it and stakeholders keep demanding it, mandatory ESG reporting could end up in the same quarterly reports that contain a company's accounting statements, and there will be no more split personality as we saw between Goldman's ESG and shareholder reports.

But some pioneers are not waiting for the government to act. Take Philips, the Dutch health care technology company. Though once an industrial conglomerate (many consumers still associate the brand with light bulbs), Philips now focuses on selling MRI machines, CT scanners, and other medical products. In the process, it's become a world leader in sustainability under its CEO, Frans van Houten. We now know to take all ESG rankings with a grain of salt, but as of October 2019, CSRHub, which aggregates rankings across dozens of sources, put Philips in the ninety-eighth percentile of all companies globally for corporate social responsibility. That's even ahead of Unilever's ninety-fourth percentile. Among other accolades, in 2019, *Fortune* magazine ranked Philips first in the world for sustainability.

Philips has achieved this by holding itself accountable: It sets specific, measurable goals that are core to the business. It frames them around the widely recognized UN Sustainable Development Goals (SDGs), focusing on three of them. First, to ensure healthy lives and promote well-being for all at all ages. Second, to ensure sustainable consumption and production patterns. And third, to take urgent action to combat climate change and its impacts. Philips breaks each one down into specific goals that it will hold itself accountable to. For example, it has committed to becoming carbon neutral in its operations with 100 percent of its power coming from renewable sources by the end of 2020.

To ensure sustainable consumption, Philips also committed to joining the circular economy. By the end of 2020, it will close the loop on all large medical equipment: when a hospital replaces an MRI machine, Philips will take back the old one to reuse, repurpose, or recycle the parts. The company is committed to doing the same, by 2025, for *all* medical equipment it sells.

Philips publicly reports on its progress quarterly, including in its annual report to shareholders. It spends nearly as many pages outlining its social and environmental performance as it does on detailing its financial performance. And it has the results audited by an outside firm.

It's important for Philips to create this accountability, because achieving these goals will require significant, sustained investment. "Short term, there is a cost of fulfilling all these targets we set for ourselves," said Edward Walsh, who sits on Philips' sustainability board and is responsible for its sustainability reporting. Committing to the circular economy, for example, is expensive. But in the long term, Philips sees these efforts as more than paying off. "Sustainable business and our vision is an enormous advantage for us," said Walsh.

He pointed out how important this deeper purpose is in recruiting, retaining, and motivating employees, especially as Philips underwent a significant transformation over the past decade. "When you drive so much change, you can lose people along the way," he said. "But one of the main constants is everybody believes in our vision and mission. And that has been a big asset for us during this relentless change."

The pace of change is only accelerating. Tony Davis had spent many years at Goldman Sachs before cofounding Anchorage Capital, a $30 billion investment firm. For most of his career, he was skeptical of ESG. He always saw it as more of a PR exercise than something real or fundamental. But after leaving Anchorage, he started doing some personal investing in small, mission-driven businesses and saw how a focus on important ESG matters could build better businesses. "I'd become a full convert," he told us.

After he launched a new investment firm a few years ago that would fully integrate ESG into its strategy, many of his industry peers were skeptical. "When I told people I was going to do this—especially some of my old private equity friends—they thought I was crazy." But recently, investment banks have been inviting him to speak to their research divisions about how to integrate ESG into their analysis. "It's incumbent on all of us that it isn't just a 'check the box' exercise," Davis says. He echoes many of the themes we've examined in this chapter: the necessity for standardized measurement and mandatory disclosures. For its part, Philips would welcome this sort of standardization and regulation in reporting. It knows it would fare well against the competition. "A lot of people talk a good talk," said Ed Walsh, "but there are a lot of words out there and not enough commitment."

DOES IT PAY TO BE GOOD?

Peter Thiel—the billionaire we met earlier who bought a panic house in New Zealand—has his own take on corporate social responsibility.

Imagine you're running a restaurant in Silicon Valley. It's competitive. You're fighting just to survive. You can't afford to pay workers more than minimum wage or worry about the environment. Some of the people who eat there work for Google. Google's market share in online search is over 90 percent. Without competition, Google can spend all it wants on its various stakeholders. Google's corporate social responsibility is "characteristic of a kind of business that's successful enough to take ethics seriously without jeopardizing its own existence," Thiel wrote. CSR, in other words, is a luxury good.

Thiel's theory was on display in 2019, when two corporate leaders were having very different Octobers. Marc Benioff, the founder of Salesforce, was publishing a book about "the power of business as the greatest platform for change." Meanwhile, Mary Barra, the CEO of General Motors, was negotiating with the United Auto Workers to end a monthlong strike that was costing the automaker billions of dollars.

Does anyone think Mary Barra would rather have been battling with workers over plant closures than preaching the gospel of stakeholder capitalism? In Thiel's framework, GM is the restaurant; Salesforce is Google. Though GM has nearly ten times as much revenue as Salesforce, Salesforce's market cap is almost triple GM's. If Benioff argues that corporations should treat their workforces better, we might expect Barra to reply, "Easy for you to say." So is Thiel right? Is corporate social responsibility a privilege affordable only by the most successful companies? Throughout this book, we've seen examples of companies that seem to fare better or worse based on how they treat their stakeholders over the long term. But as the saying goes, "In God we trust; all others bring data." Let's look at the data.

In 2015, in one of the most exhaustive analyses to date, researchers in Germany aggregated the results of 2,250 individual studies that compared companies' environmental, social, and governance criteria with their financial performance since the 1970s. The researchers found that 63 percent of the studies show a positive relationship between ESG and financial performance and only 10 percent of the studies show a negative relationship. These results are impressive, given how noisy the ESG data can be. And they held regardless of region, time period, and asset class.

Other researchers found that what really mattered was a company's ESG rating on issues that were *core* to the company's industry. Financial risk management matters in finance but not in transportation. Carbon emissions matter in transportation but not in finance. (May Lehman Brothers rest in peace.) Researchers found that outperforming on unrelated ESG metrics did not improve financial results, but that outperforming on core ESG metrics did. Essentially, doing good in unrelated, marginal ways doesn't pay. But doing good in the core of your company does. This should be encouraging for any company that is making ESG central to its business model.

In the last study we'll highlight, researchers compared a company's sustainability policies in 1993 to how it performed in 2009. Many studies compare ESG and financial performance over several quarters or a few years. This one compared them over a decade and a half. The long time frame allowed researchers to examine how responsibility and return interrelate. After analyzing the data, they found "evidence that *High Sustainability* companies significantly outperform their counterparts over the long-term, both in terms of stock market as well as accounting performance."

The metrics aren't perfect. The data aren't perfect. But the studies that exist suggest that the anecdotal patterns we've written about are part of a broader truth: companies that serve stakeholders well usually serve shareholders well, too.

Why should this be true? A social purpose can clarify strategy, as it helped CVS and AXA to stop doing business with tobacco. A meaningful mission can attract and motivate good workers, as it did at Philips during its transition from an industrial conglomerate to a health care company. It can lead companies to think longer term and make investments in sustainability that eventually reduce costs. It can reduce the long-term risk of legal action, regulation, boycott, or protest.

Or we can boil it down to common sense: When you invest more in stakeholders, they'll invest more in you. When you genuinely care about serving others, they'll help you succeed. Successful corporate strategy depends on countless interactions among a large number of different parties, including consumers, employees, competitors, regulators, activists, the press, and communities. We try to describe corporations and strategy in quantifiable metrics—ESG, financial, operational—but the map is not the territory. In a competitive economy, trust, commitment, and goodwill are the intangible lifeblood of long-term success.

And about that restaurant in Silicon Valley: Thiel is right; it's a competitive business. But today, restaurants are often competing on which can offer the most locally sourced, sustainable fare. The salad restaurant MIXT has nine locations in San Francisco alone—including one just a seven-minute walk from Google's San Francisco offices. MIXT literally paints its impact on its walls: four out of five store managers are promoted from within. The average employee is with the company for three years, compared with a few months for most restaurants. It diverts 99 percent of the waste it generates away from landfills. It's a Certified B Corporation. So, yes, restaurants are a competitive business. This is how MIXT competes.

A FITTING CODA

For too many, corporate social responsibility and its ilk—ESG ratings, sustainability reports—remain a superficial marketing exercise rather than a fundamental change in strategy. Today's efforts are often well intentioned. But without accountability, they won't amount to real and lasting change. We are sympathetic to the words of Paul Adler, a UCLA economist who is skeptical of hollow reform: "When the patient has cancer and needs major surgery, dieting is nice but not a cure, and it is dangerous to encourage the patient to think otherwise."

CSR is at its best when it addresses the core impact that corporations have. Its measurement should be standardized and reported in mandatory, audited statements, right beside its financial results. This is how we begin to create the accountability required to achieve real change.

Indeed, even Larry Fink, for all his corporate purpose proselytizing, has come under criticism for failing to line up his actions with his rhetoric. He was protested at the Museum of Modern Art in New York City, where he is on the board, because of BlackRock's investments in private prisons. He was protested at BlackRock's annual shareholder meeting for not taking big enough steps on climate change. BlackRock has already offered strong public support for standardized metrics such as SASB, but people are demanding more.

Society's expectations of corporations have changed over time, but corporations have failed to keep up. Consumers and employers are expecting more from the corporations they interact with. And so are some shareholders.

That last point is significant, because, as we've seen, shareholders have the final say in our capitalist economy. This is the Two-Buffett Paradox that we saw in chapter 1: when stakeholders and shareholders butt heads, CEOs too often respond with rational hypocrisy. Making fundamental change the way Unilever, AXA, and Philips have done requires shareholder buy-in to be successful and sustained over time.

Let's look again at the Business Roundtable. The Business Roundtable has proven itself a bellwether of capitalist sentiment. After its previous rotation on purpose and profit, it turned around again in 2019. It released a statement redefining the purpose of the corporation as being to promote an economy that serves all Americans. It once again listed shareholders last, re-recognizing that building valuable businesses in the long term requires recommitting to a purpose deeper than profit.

There's just one problem: in capitalism, the capitalist is king. In response to the Roundtable's new statement, the Council of Institutional Investors released its own. The Council represents members managing $39 trillion of assets. If the Roundtable runs our corporations, it does so only at the pleasure of the Council. As the *Wall Street Journal* editorial board admonished, "CEOs are themselves employees hired by directors who are supposed to be stewards of the capital that shareholders have invested." The capital holders were ready to speak.

Against the Roundtable's vision of stakeholder capitalism, the Council wrote, "Accountability to everyone means accountability to no one." It reminded the Roundtable that corporations need to keep their focus on shareholder value and leave the rest to government. A corporation's duty—its fiduciary duty—is to maximize profits for shareholders. We've heard that one before.

The Watergate scandal taught us to follow the money. When we pull the thread of capitalism, it leads first to corporations blinded by fiduciary absolutism. Keep on pulling, and we find that even CEOs answer to someone: their institutional investors. Pension funds, labor unions, and investment companies manage the assets, but they don't manage them for their own benefit; they manage them for *ours.* We are the retirees and savers and union members for whom the Council of Institutional Investors ultimately speaks. We've pulled and pulled and pulled and found that the thread leads directly to our own pockets.

We talked about Dr. John Harvey Kellogg and his brother Will as the Dr. Jekyll and Mr. Hyde of our corporate world, dueling between purpose and profit. But as Robert Louis Stevenson wrote it, Jekyll and Hyde are not two people but one. Hyde suffers from a split personality: during the day he is an upstanding citizen; at night he turns into a terrible beast. Over time he is overwhelmed by the murderous Hyde.

Pull the thread on our economy, and we see it's not a tug-of-war between good guys and bad guys, between the moral and the maniacs. It's just us. In the Two-Buffett Paradox, *we are both Buffetts.* We demand that corporations be more responsible, and then the institutional investors that represent us demand that they keep maximizing profit. Just as we think we've finally met the villain face-to-face, we find we're staring at a mirror.

In this chapter, we focused on corporations and the ways they are trying to change for good. But we've now seen how essential it is for investors to be on board, if not leading the change. In the next two chapters, we will show how responsible investors are trying to do just that. We'll start with one of the most contentious battles over responsible capitalism today: divestment. It's a fight that has pitted two sides against each other—university students against administrators—who couldn't agree more on the problem of climate change but couldn't agree less on what to do about it.

5

HEAR NO EVIL, SEE NO EVIL

DIVESTMENT, ENGAGEMENT, AND AN EXISTENTIAL CRISIS

Lawrence Bacow, still in his first year as the president of Harvard University, could barely be heard above the protestors' chants. It was the spring of 2019. He stood at the front of the stage at an event for the university's Kennedy School, a microphone clipped to his yellow tie as he mouthed remonstrations above continuous shouts of "Disclose, divest, or this movement will not rest."

"You're not being helpful to your cause and I suspect you're also not gaining many friends or many allies in the audience by virtue in the way in which you choose to express your point of view," he told them.

Meanwhile, students in the crowd unfurled a massive banner from the balcony that read DIVEST HARVARD above a clenched fist. Six activists had joined Bacow on the stage and sat silently holding signs demanding that Harvard divest its endowment from a number of morally fraught industries, fossil fuels first among them. One had replaced the "Veritas" on Harvard's crest with the word "Slavery." The Divest Harvard campaign had shown up in full force.

The students wanted to know: How could an institution supposedly committed to combating climate change willingly invest part of its $39 billion endowment in oil and gas companies? According to a university spokesperson, Harvard fights climate change through "research, education, community engagement, dramatically reducing its own carbon footprint, and using our campus as a test bed for piloting and proving solutions." Yet when it comes to its endowment, Harvard owns the very companies that generate significant environmental harm.

Harvard's official line on divestment is clear:

> The University maintains a strong presumption against divesting itself of securities for reasons unrelated to investment purposes, and against using divestment as a political tool or a "weapon against injustice"—not because there are not many worthy political causes or deeply troubling injustices in the world, but because the University is first and foremost an academic institution.

Divest Harvard takes the opposite stand: "We condemn investment in the fossil fuel industry for its role in actively perpetuating the climate crisis." Divest Harvard demands that the university commit to divesting the endowment from all direct and indirect holdings in the fossil fuel industry, and then use its financial power to advance "reparative justice by supporting environmentally sustainable, socially responsible, and community-based investment."

"The faculty and students who are in this now are in it for the long game," said Ned Hall, a Harvard philosophy professor helping to lead the faculty's divestment push. "We're not going to be deterred. We're not going to stop." One student leader focused on how high the stakes are: "Climate change is the greatest injustice of the century."

Decades before Divest Harvard began, a team of scientists from ExxonMobil achieved something remarkable: they wrote a memo, recently uncovered during a Pulitzer Prize–winning investigation, that predicted how the growth in atmospheric carbon dioxide would increase global temperature over time. The memo included a chart that, now grainy from reproduction, projects the path of climate change almost exactly. In 1982, based on the expected increase in fossil fuel use, they correctly predicted that we would hit a carbon dioxide concentration of 415 parts per million and an increase in global temperatures of 1.8 degrees Fahrenheit.

The company's own scientists warned that we "would warm the earth's surface causing changes in climate affecting atmospheric and ocean temperatures, rainfall patterns, soil moisture, and over centuries potentially melting the polar ice caps."

Climate change is creating an existential crisis for our planet. Meanwhile, global energy consumption is projected to grow another 28 percent within a generation. How could Harvard, an institution whose motto is Latin for "Truth," stand to invest in companies such as ExxonMobil that actively conceal the truth about the consequences of their business?

Bacow is not the only university president facing calls for divestment; Harvard is only the highest profile of dozens of institutions where the divest movement has taken hold. Around the world, concerned constituents are protesting their institutions' ties to the oil and gas industry—from the Anglican Church to the New York City Pension Fund, from Stanford's endowment to the Norwegian sovereign wealth fund. Never mind that Norway's trillion-dollar sovereign wealth fund was funded by oil money in the first place. The Rockefeller family charity also atones for its sins by divesting from the very industry that enriched it. Today, more than one thousand institutions managing $11 trillion of assets worldwide have divested themselves from oil and gas companies. That's up from $52 billion in 2014.

There is a clear logic to the divest movement. Ownership entails responsibility; if you own shares of a company that is destroying our planet, you are responsible for that destruction. The movement calls attention to the hypocrisy of an institution that fights climate change with one hand while propagating it with the other. It busts the myth that there is such a thing as "value-neutral" investing. And in its stead, divestment offers a change that is simple, tangible, and achievable: sell your sin stocks. Absolve yourself. Despite the institutional intransigence the movement faces, divestment is satisfyingly obvious.

Just one question before we join the march: Does it work?

A CALL TO ACTION

"I was twelve," remembered Ilana Cohen, one of the student leaders of Divest Harvard. "It was in October of 2012. I would have been in seventh grade." She, like many Divest Harvard students, traced her activism to a natural disaster. For her, it was Superstorm Sandy hitting New York. "I was in my apartment for eight days," she said.

One thing that separates Gen Z and younger millennials from older generations is the narrative that accompanies their first experience of natural disaster. For most adults, their first natural disaster was understood as only that—a natural disaster. For younger adults today, the experience is inextricably tied to climate change. Twelve-year-olds in 2005 learned that Hurricane Katrina was a devastating act of nature; twelve-year-olds in 2012 learned that Superstorm Sandy was a devastating consequence of human indifference. "For someone growing up in California now," said Galen Hall, a climate activist at Brown University and Professor Hall's son, "their entire life and landscape is shaped by climate change."

For Cohen, it was not her or her family's suffering that got her engaged in the movement. It was how the storm had hurt other communities less privileged than her own. Her school opened within days after the storm, but other schools were closed for weeks. "All of the areas of midtown Manhattan recovered quickly, whereas these areas on the outskirts—disproportionately lower-income communities and communities of color—felt these effects for a much longer time," she said.

When Divest Harvard fights, it is fighting for marginalized populations. It's fighting for climate justice. "I think that environmentalism and social justice are fundamentally inseparable," Cohen said. "There are no climate change issues that are divorced from social justice."

The modern divest movement began not with climate but with apartheid. In the late 1970s, black South Africans continued to suffer under the oppression of institutionalized segregation. They were restricted in where they could work and live, where they could go to school, and whom they could marry.

Outraged by these atrocities, citizens in Europe and the United States sought justice. They landed on divestment. South Africa was deeply dependent on trade and foreign investment. Most of its companies were controlled by white South Africans. The idea was that divestment would put economic pressure on those companies and the South African economy overall. The logic is no different from the justification for the Cuban embargo or economic sanctions on Iran and North Korea. Economic pressure, coupled with political pressure and social demonstrations, activists felt, could help force the government to change.

At the time, Harvard students carried signs reading END HARVARD SUPPORT OF APARTHEID and PUT APARTHEID OUT OF BUSINESS. They successfully forced the university to pull investments from some companies that dealt directly with the apartheid government, though in Harvard's case, divestment was only partial and came only after many others had divested. "Harvard is always at the vanguard of the rear guard," remarked Professor Hall.

"It's not really about leadership," said Cohen, pointing out how many other institutions have already divested from oil and gas. "It's about taking responsibility and reclaiming some sort of agency in a crisis where Harvard does have the potential to play a meaningful role."

Divestment offers action on a crisis many feel helpless to prevent. It has a deeply righteous ethos and a righteous history. It gives us a momentum we lack elsewhere, because it puts a tangible victory within reach. "I actually think that now is the moment that Harvard can make the right choice and put itself on the right side of history," said Cohen. Or it could go the other way. The important thing is to fight.

Caleb Schwartz, another student leader of Divest Harvard, felt his position as a Harvard student as a call to action. "I've always had this underlying feeling of benefiting from so much privilege here," he said. "What am I doing to make the system more fair?"

INSTITUTIONAL INTEGRITY

Harvard president Bacow toes a traditional line on the purpose of the endowment. "The endowment exists to support the institution, to support our students, and to support our faculty," he told the *Harvard Crimson.* Donors give funds on those terms, not to accomplish some other goal. The "About" page of the endowment's website is bannered with the phrase "A Singular Mission." Below, it states that its mission is to ensure that the university "has the financial resources to confidently maintain and expand its leadership in education and research for future generations." This is its *singular* mission. No double bottom line or dual mission here.

Harvard is not alone in setting its endowment aside as a value-neutral pool of capital meant to provide funding for its operations. Stanford has taken a similar tack. The Stanford Board of Trustees' Statement on Investment Responsibility reads: "Just as the University does not take positions on partisan or political issues, the Trustees maintain a strong presumption against using the endowment as an instrument to advance any particular social or political agenda."

Hold on. The university doesn't take positions on partisan or political issues? What about Stanford's commitment to progress on climate change, in which the university commits to "slow, and ultimately prevent, the rise in the global average temperature"? What about the joint op-ed on climate change by the presidents of Stanford and Harvard in the *Huffington Post*? What about Stanford's public advocacy of immigration policies such as the DREAM Act? It strains credulity to insist that universities such as Stanford and Harvard are apolitical. And it would be a vastly deflated vision of our universities if we didn't believe that they serve some greater social purpose.

There are obvious ways in which universities achieve their mission: educating students; funding research and producing scholarship; establishing institutional moral leadership. But it seems strange to ignore one of their most significant sources of impact: their endowments. Harvard's operating expenses in 2018 were $5 billion. Its endowment at the time was $39 billion. Just for the sake of argument, let's assume the endowment was invested entirely in stocks. Based on market averages, $39 billion could buy a set of companies with total revenue of $18 billion and total costs of $17 billion. Compared to its $5 billion of operating expenses, 3.4 times as much spending happens through companies in which Harvard is invested as at Harvard itself. If your university has a mission, how can you ignore over three-quarters of your economic power?

This is an overstatement, both in magnitude and in how much effective control Harvard has over the companies in which it invests. But if Harvard has some set of values, some broader mission in the world, why should it care about those values and that mission in every aspect of its organization but one?

Unfortunately, social responsibility is not what endowment managers sign up for. Trained at investment banks and traditional asset management firms, they are conditioned to focus only on financial performance. They are judged by their peers on the returns they produce each year. Leading the charge on climate change and other social issues is anathema to both the professional culture and the conventional understanding of the purpose of an endowment.

The concept of "value-neutral" investing is a convenient myth. It lets endowments focus on asset allocation and manager selection and ignore social and environmental impact. But it's just that, a myth. If Harvard owned shares in Gannett, it had to vote on whether to support Gannett's incumbents or those proposed by Alden and Digital First. If it owned shares in Lehman Brothers or Goldman Sachs, it had to elect their board members. If it owned shares in Unilever, it would have had to vote on 3G's proposed takeover had Buffett not stood down.

There is no value-neutral approach in these or any other cases. There is only hiding behind the threadbare theories of the efficient market and the invisible hand, the idea that somehow maximizing share price within the bounds of the law is a default from which any deviation must be justified. Every decision—even whether to default to the status quo—is a value-laden decision.

"At the end of the day," said Schwartz, the Divest Harvard student leader, "it's the 'fiduciary duty argument' they keep coming back to, to keeping funds up for Harvard's educational mission. The moral arguments are there, but they are totally trumped by fiduciary duty." He thought for a second. "Anytime they talk about duty, it's always about maximizing return. Sustainability is not a duty they feel."

Compare this to a foundation such as the United Kingdom's Joseph Rowntree Charitable Trust. The trust was established in 1904 as "a Quaker trust which supports people who address the root causes of conflict and injustice." Its founder, the nineteenth-century chocolatier Joseph Rowntree, believed that "for your efforts to have any lasting benefit, you must tackle the roots of a problem" with every tool available. Otherwise, you will only ease the symptoms without making a lasting difference.

Unlike Harvard, the trust sees no separation between what it supports through its programs and its endowment. It acknowledges that its investments "do not stand in isolation from our grant giving work" and that those investments are a key part of achieving the trust's overall mission.

The full integration of an organization's values into its endowment would represent the greatest realization of divestment's vision. To traditional endowment managers, doing so would precipitate a descent down a slippery slope. After oil and gas companies, what would be next? Defense companies? Companies that produce unhealthy food? For-profit education companies? Any company that doesn't pay a $15 minimum wage?

The divest movement forces us all to recognize that there is no value-neutral way to invest. The companies Harvard owns, the way it votes those shares, the managers it selects—they all reflect a set of values, whether Harvard acknowledges it or not. As Cohen, who was spending an early-fall morning protesting outside Bacow's residence, put it, "We can't just go about our lives going to class every single day. We can't just be university presidents living in our very nice houses and going straight into our offices. . . . We all have to take responsibility as active participants in building a more just and stable society." And that includes the $39 billion in Harvard's endowment.

FLEEING THE BATTLEFIELD

Once we accept that our investments will always reflect a set of values, there's still the question of how best to make our values heard. Divestment is one way, but it's not the only one.

The sorry story of NRG, one of the country's largest power producers, pushes us to consider whether divestment is the right one. NRG suffers from sustainability whiplash. The public company sells electricity across the country, with nearly a hundred power plants in eighteen states. Under its former CEO David Crane, the company set aggressive goals to lead the way in combating climate change. "The day is coming," wrote Crane in a 2014 letter to shareholders, "when our children sit us down in our dotage, look us straight in the eye, with an acute sense of betrayal and disappointment in theirs, and whisper to us, 'You *knew* . . . and you didn't do anything about it. *Why?*'"

So Crane led NRG to publicly announce that it would cut its carbon dioxide emissions by 50 percent by 2030 and 90 percent by 2050. It made massive investments in renewable energy, particularly wind and solar. It looked decades into the future to figure out a way to be a leading, prosperous, and sustainable power company.

But not everyone was excited by those commitments. Like all public companies, NRG is subject to the demands of its shareholders—including activist hedge funds such as Paul Singer's Elliott Management. Unhappy with NRG's financial performance and unimpressed with its sustainability goals, in 2017, Elliott announced that it had bought a large block of shares and had plans to transform NRG back into a traditional carbon-based power company.

As one of the largest shareholders, Elliott was soon able to name two new members to NRG's board, including Barry Smitherman, a former Texas energy regulator who once called climate change a hoax and gave a presentation titled "The Myth of Carbon Pollution." With Elliott now wielding influence in the boardroom, NRG released a plan to restructure the company by selling off its renewable energy assets. The board forced out Crane as CEO, ending his twelve years of leadership at the company.

Elliott was able to force change at NRG because it owned part of the company. That's how ownership works in a capitalist economy. But Elliott didn't own the whole company. The fund owned only 6.9 percent of the shares. Even with its partner—the private investor Bluescape Energy Partners—it could speak for only 9.4 percent of the ownership. Where were the other 90.6 percent of shareholders? Where were all the shareholders who cared about climate change, about the long-term viability of carbon power, about the need to transform our electrical grid? Why didn't they speak up, supporting Crane and forcing Elliott to back off? Where were universities such as Harvard, with their firm commitments to combating climate change, their vast networks of connections, and the moral stature to rally others to their cause?

Shareholders who aren't paying attention or choose to stay silent make it possible for activist hedge funds such as Elliott to have their way. But you can't have a say if you're not even at the table. Every day, corporations make decisions that will determine our future. The responsibility for these decisions ultimately falls on shareholders. At NRG and elsewhere, we need engaged investors such as Harvard fighting to build companies that reflect our values over the long term. We need good guys in the room. But at our most perilous companies in our most morally fraught industries, divestment explicitly and deliberately clears the room of our most responsible investors. The ones who care the most are the ones who aren't there.

David Crane had to watch as Elliott stormed the company and undid what he had done. Reflecting on his legacy, he offered a sobering lesson. When Elliott came knocking, "there was no institutional investor support for NRG's attempt to go green." There's all this "happy talk coming out of the senior ranks of the major pension funds, sovereign wealth funds and university endowments about investing their money in a climate positive way," but when it came time to make hard choices, he found only "money managers who are, at best, climate-indifferent."

THE CASE AGAINST DIVESTMENT

In examining the impact of the anti-apartheid movement on South Africa's economy, most researchers point out that boycotts had a far greater impact than divestment did. International boycotts of companies owned by rich white South Africans had an immediate impact on their profitability as their sales plummeted. The cause and effect were clear.

But divestment is a different beast, and the effects are not so obvious. For one investor to sell a share, another must buy it. Period. An unhappy shopper can complain to management or take his business elsewhere, but an unhappy shareholder simply replaces herself with someone who cares less.

If Harvard were to sell all its shares in oil and gas companies, what would happen to the oil and gas companies' operations as a result? Well, nothing. All that would have changed would be that Harvard used to own some of their shares and now someone else owns them instead. Selling shares to another willing owner does not have a direct impact on the underlying profitability of a company or an industry. It doesn't change the cost of drilling for oil, refining oil, transporting oil, or burning oil. And it doesn't change the cost of the alternatives.

There are two arguments in favor of divestment that are worth addressing. The first is that divestment makes it harder for corporations to raise capital and operate. The second is that even if it doesn't, at least it keeps us from profiting off an industry that we think is harmful. Neither is as straightforward as we'd hope.

On the first, proponents argue that divestment should make it harder for oil and gas companies to raise capital. With fewer investors willing to hold their shares, the share price will fall and it will become more expensive for them to raise money to invest in carbon-intensive projects. Think about trying to get a loan: If there are a dozen banks competing to be your lender, you can expect to get a very competitive interest rate. If eleven of them refuse to do business with you for whatever reason, you'll end up paying a higher interest rate to whichever lender remains. In finance, we call this interest rate your "cost of capital." Proponents argue that divestment is effective because it raises the cost of capital for oil and gas companies.

Even if divestment does successfully raise a company's cost of capital, it's an extremely indirect mechanism for social change. A boycott immediately and measurably inflicts financial pain on the offending company. Divestment hypothetically increases the cost of capital in some future state when the company is trying to raise new money.

And how often are our public companies raising new capital anyway, given that many can finance new projects out of their existing cash and profits? Not often. Over the last decade, public companies have raised a net *negative* $287 billion from shareholders. That is, $287 billion more has been doled out to shareholders than raised from them. In our loan example, divestment is like eleven banks refusing to give you a loan when you've already had one for a decade and aren't looking for a new one. Companies like NRG and ExxonMobil do not need our money; they need our votes.

In an interview with the *Financial Times*, Bill Gates was blunt in his assessment of the divest movement's impact: "Divestment, to date, probably has reduced about zero tons of emissions. It's not like you've capital-starved [the] people making steel and gasoline." Gates is supported by economists such as Tyler Hansen and Robert Pollin, who published a paper in 2018 assessing the impact of the fossil fuel divest movement. They concluded that "divestment campaigns, considered on their own, have not been especially effective as a means of significantly reducing CO_2 emissions, and they are not likely to become more effective over time." In an interview,

Pollin noted, "Most efforts now devoted to divestment campaigns would be better spent on more direct efforts to drive down fossil fuel consumption."

In 2015, both *The New Yorker* and *The New Republic* ran articles on the same question: Does divestment work?

William MacAskill, one of the founders of the effective altruism movement, summarized the existing research for *The New Yorker*. "If the aim of divestment campaigns is to reduce companies' profitability by directly reducing their share prices, then these campaigns are misguided." Divestment's effects on industries such as gambling and tobacco "suggest they have little or no direct impact on share prices." In her analysis for *The New Republic*, the journalist Rebecca Leber concluded, "It's near-impossible to prove that divestment alone has ever had any financial bearing on their targets." Looking at both the anti-apartheid and antitobacco movements, she added, "Both are considered a success—though not because they financially impacted their targets. They didn't."

In some cases, divestment might even result in the unintended consequence of enriching less moral investors. When New York City's five public pension funds decided, in 2018, to divest their $5 billion from fossil fuels, some portion of their shares were likely bought by less scrupulous investors. There are private investment managers like Dan Ahrens, who focuses exclusively on "sin stocks." For decades, he's acquired shares in gambling, alcohol, and firearms companies—the very same companies that others attempt to blacklist. Ahrens thinks he can outperform the rest of the market by focusing on stocks that other investors avoid.

Sin stock investors such as Ahrens may actually make *more* money thanks to the divestors. Let's assume that so many investors divest that the cost of capital for an oil and gas company doubles. When the cost of capital doubles, the expected return for an investor in that company—the person to whom this "cost" is being paid—must double as well. Think again about the interest rate you pay on a loan. Higher interest expense for you means higher interest income for your bank.

The more successful divestment is, the more money there is to be made by investing in the shunned companies. As long as nothing changes about the underlying economics of oil and gas production, the divest movement has the perverse effect of enriching investors who do not care about the moral causes divestment takes on. If divestment were successful enough to start increasing the cost of capital, do you think investors would let this higher expected return pass unnoticed?

What if so many investors divested that no upstanding public investor dared invest? Would companies fold and the industry cease to exist? The porn industry proves that that's not how it works. You don't hear about a "Divest Porn" movement. Why? There are no public porn companies. That doesn't mean there's no pornography. Indeed, porn accounts for 25 percent of all internet searches and 35 percent of all downloads and makes up three of the top thirteen most trafficked sites. But porn companies are all owned by private investors, who are happy to run the industry outside the public eye. Divestment doesn't change the underlying economics of an industry; it just concentrates the ownership of our most morally fraught companies in the hands of our least scrupulous investors.

Imagine that a group of kids finds a gun in the woods. They're walking to turn it in at the police station. Who should hold the gun? It should be the most responsible kid, the one who understands how guns work and the dangers they pose, right? Based on the tenets of the divest movement, the most responsible kid would refuse to touch the gun on principle.

Ownership is power. In our capitalist economy, the capitalist has the final say. If Harvard is a partial owner of ExxonMobil, it is indeed complicit in everything ExxonMobil does. But buying and selling shares in a company on a stock exchange merely changes who holds an ownership stake. As divestment advocates stress, Harvard is indeed responsible for the economic power it wields through its endowment. But divestment is an abdication of that responsibility, not its fulfillment.

But what about the second argument for divestment—that we should at least refuse to profit off an immoral industry on principle?

For this one, it's useful to return to divestment's roots. Though the modern divest movement traces its lineage back to the boycotting of South Africa during the era of apartheid, its true genesis was the refusal of Quakers to own companies connected to the slave trade. Quakers were not opposed to business or capitalism. Indeed, they amassed some of the largest fortunes in the early Americas. But their faith precluded them from profiting off certain activities. They would neither gamble nor profit from gambling. They would neither fight in wars nor finance them.

This approach allowed them to preserve their righteousness in business. It represents the "hear no evil, see no evil" philosophy of divestment, an approach shaped and informed by religious conceptions of sin and moral purity.

Divestors claim that they're refusing to profit off of oil and gas, but what about the money they receive for selling their shares? The price of a share today should approximate the present value of all its future cash flows. If markets are even close to efficient, then these two things—receiving a dividend check every year forever or receiving the stock price today—are roughly the same thing economically.

We wind up with a near-comical result: Divestment doesn't even mean that divestors don't receive profits from oil and gas companies; it just means they are cashing out future profits today. Divestment does not reduce the amount of profit an endowment makes from oil and gas companies; it just changes *when* the endowment will receive that profit. It's as if protestors are demanding "Don't profit from oil in the future—profit from oil today!"

THE CASE FOR ENGAGEMENT

Where does that leave us? If we believe that institutions must reflect their values in their investments but that divestment doesn't achieve its aims, what should they do instead?

There are several promising ideas. They require investors not to wash their hands of a company but to roll up their sleeves—to engage with the companies they own and reform them from the inside.

In 2019, Monster Beverage, the massive purveyor of highly caffeinated energy drinks, scored dead last on a list of companies ranked by how responsibly they managed their supply chains. For Andy Behar, the leader of the shareholder advocacy organization As You Sow, Monster's indifference to its supply chain was unacceptable. Through As You Sow, Behar organizes existing shareholders to use their power as owners to further specific social or environmental issues.

When he first contacted Monster, the company brushed him off. Undeterred, he escalated his request to a shareholder resolution that would be voted on at Monster's annual shareholder meeting. He helped organize a petition with 21,000 signatures from customers and rallied the support of some of Monster's largest suppliers, including Coca-Cola. If human rights in its supply chain had not been a top issue for Monster before, it was now.

Sure enough, Monster reacted. Within a year, it had audited 80 percent of its suppliers. The company now works hand in hand with As You Sow and other nonprofits to ensure that it's following best practices in supply chain transparency. With engaged and active shareholders leading the charge, Monster had no choice but to shape up.

The SEC states that any shareholder who holds either $2,000 or 1 percent of any company's stock for at least a year can submit a proposal for a vote at the company's annual meeting. Large investors have the ability to engage with their companies between meetings and beyond specific votes. Norges Bank Investment Management, Norway's sovereign wealth fund, uses engagement to push for its climate change agenda. In 2016, this resulted in nearly two thousand meetings in which there was discussion of environmental, social, or governance issues. Other institutional investors such as Fidelity are beginning to change their policies so that they can support shareholder proposals "calling for reports on sustainability, renewable energy, and environmental impact issues."

The California Public Employees' Retirement System, known as CalPERS, scored a significant win against the prevailing practices at the oil giant Occidental Petroleum Corporation. CalPERS manages $320 billion worth of pension funds for 1.8 million retired California public employees. It submitted a shareholder resolution in 2017 that would require Occidental to explicitly account for climate change in its strategy, policies, and operations. The passage of this resolution over protest by Occidental's board marked the first time that a climate-related shareholder-led resolution was successful at a major US oil and gas company.

That same year, a similar "shareholder rebellion" took hold at ExxonMobil. Nearly two-thirds of shares supported a resolution that would require ExxonMobil to report on the impact of climate change, against management's wishes. As with Monster and Occidental, ExxonMobil ultimately bowed to investor pressure.

Divestment is symbolic, and symbols matter. But we cannot settle for symbolic victories. Divestment has us vote with our feet—with questionable impact—but engagement has us vote with our vote.

REEVALUATING ENGAGEMENT

Nonetheless, engagement has its skeptics. "It's just mind-boggling that people take this seriously," said Ned Hall, the Harvard professor helping to lead the divest movement. In his own assessment of Harvard's record, engagement has amounted to "some extremely minor token victories."

Student leaders of the Divest Harvard movement argue that engagement is insufficient because, small victories notwithstanding, oil and gas companies are still perpetuating environmental harm.

Andy Behar, who used engagement to force Monster to change, has been blunt in his own assessment of engagement at oil and gas companies. Of the 160 climate resolutions filed at twenty-four oil and gas companies in the United States since 2012, he believes that none has led to material change, those at Occidental and ExxonMobil included. In his assessment, engagement has failed to accomplish anything beyond superficial changes to corporate policies.

Divest Harvard remains skeptical that these corporations will ever make the transformative changes necessary to fight climate change. "I don't think my generation really has faith in these big companies to get there fast enough," said Caleb Schwartz, a student leader of the movement. Divestment is both the last resort and the only option.

Hardly idealistic or naive, the students involved with Divest Harvard are familiar with every argument against divestment. "We're aware of the fact that Harvard divesting is not going to bankrupt companies," said Schwartz. "We know somebody's going to buy up those shares—possibly someone with less good intentions."

So why divest? For them, divestment is the first step in massive systemic change. Rather than looking at the proximate effects of divestment within the bounds of our existing market and laws, Divest Harvard has its eye on a more distant horizon. It's "much more than just market supply and demand," Schwartz said.

"You can call it a political act if you want to," said Professor Hall. "You can also just call it an act of civic leadership." The idea is as follows: if enough well-placed institutions divest, the political conversation will be fundamentally altered. Systemic change will never come from within corporations; it will have to be imposed on them through political change and legal reform. As Hall says, shareholder engagement will never result in a carbon tax. "It's just not going to happen."

Student leaders such as Schwartz believe that divestment exacts a cultural toll, labeling oil and gas companies "morally repugnant" and taking away their license to operate as social actors. The divest movement is about creating such a powerful social stigma to climate injustice that fundamental change becomes possible. If big players such as Harvard say "enough is enough," Schwartz says, it will be easier to pass bold legislation rapidly and with broad political support.

This is the deeper divestment theory of change. What really matters is that Harvard talks about divesting, and that climate change is on the minds of every Harvard student and every Harvard faculty member, planting the seeds of revolutionary cultural change in a new generation of leaders.

By this logic, the divest movement ought not to measure its success by the number of endowments that ultimately divest. It ought to measure its success by its ability to center climate change in broader political discourse, to marginalize companies, to pass new laws, and to change energy consumption. In some ways, the endowment's obstinacy has actually helped increase Divest Harvard's profile. If you're going to divest, it's better to have a big public fight about it first; that will force climate change to remain at center stage much longer than if the endowment had acquiesced immediately.

But can't Harvard do both? Can't it push for systemic change and also exercise power as an active and engaged owner? "That's a good question," said Professor Hall. "If someone could make a credible case, we could do as effective a job and use our leverage as shareholders—if that could really be credibly put forward, you know, I might be convinced."

Unfortunately, there's not much faith in Harvard's engaging effectively. There's no confidence on the part of faculty or students that the endowment's efforts are even close to doing so. Instead, there's widespread suspicion that Harvard's board of trustees would prefer that the faculty and students stop trying to meddle in its endowment. "You have a lot of people who move in a world where their job is to make money and see other views as somehow naive," Professor Hall said.

STILL ASLEEP AT THE SWITCH

If the divest movement can trace its lineage back to slavery resistance, shareholder engagement can trace its way back to napalm. In 1968, the Medical Committee for Human Rights owned five shares of Dow Chemical Company stock. When it tried to submit a shareholder proposal banning the sale of napalm, Dow blocked it. The committee sued, and the courts eventually forced Dow to allow shareholders to vote.

When the vote finally occurred to ban the sale of napalm, it was nearly unanimous. Shareholders representing only 3 percent of shares supported the proposal. Napalm sales continued unabated. Three hundred eighty-eight thousand tons of napalm would be dropped on Vietnam in all.

Half a century later, shareholder engagement continues to yield lackluster results. From 2009 to 2018, there were 1,372 shareholder proposals that were environmental, social, or political in nature. However, on average each proposal was supported by only 21 percent of shareholders. Success is rare. It's hard to organize absentee owners and apathetic institutional investors blinded by fiduciary absolutism.

The resolution at ExxonMobil passed only because it had the support of large institutions such as BlackRock and Vanguard. But that same year, BlackRock and Vanguard supported only two of ninety climate change–related proposals they voted on. Though BlackRock has publicly championed environmental responsibility, it has supported 99 percent of management-nominated directors at US fossil fuel companies and has failed to support 90 percent of all climate-related shareholder resolutions.

If institutional investors are unlikely to support a proposal, they are even less likely to initiate one. A review of almost four thousand shareholder proposals from 2008 to 2017 did not surface a single proposal filed by Vanguard, BlackRock, or State Street. The late Jack Bogle, the founder of Vanguard, was quick to acknowledge the firm's passivity. "For decades, with a handful of exceptions, the participation of our institutional money managers in corporate governance has been limited, reluctant and unenthusiastic," he admitted. For 90 percent of the companies they are invested in, Vanguard, BlackRock, and State Street don't take a single meeting with management all year.

This lack of engagement is perplexing, given that these institutional investors are charged with representing the interests of workers and retirees, who care not just about returns but about living in a livable world. Sure enough, in 2019, one organization gathered 129,000 signatures to petition BlackRock and others to support climate-related proposals that are better aligned with the interests of their stakeholders.

Engagement is messy, hard to police, easy to fake. Divestment is satisfyingly simple.

At Harvard, the underlying problem is trust. Administrators so far have failed to demonstrate that their engagement will be more powerful than divestment would be. They have failed to prove that their commitments are more than empty talk.

There's a long way to go in restoring that trust. Though the endowment gives students, alumni, and faculty a voice through an advisory committee, the endowment is under no obligation to abide by the committee's recommendations.

In one attempt at transparency, the endowment released a memo about its sustainable investing. It detailed two successful instances of engagement that had pushed companies to change their strategy. But the two companies were opaquely referred to as "Company A" and "Company B." Anonymity is a long-standing practice in the world of engagement, but the statement does little to elicit the confidence of the wary audience.

Harvard's voting record on climate change also does not inspire trust. In 2018, there was a shareholder proposal at Chevron to issue a report about how the company could reduce its dependence on fossil fuels. How did Harvard vote? It abstained. In its justification, the endowment wrote:

> When considering company strategy on a core question of this kind, shareholders might prefer to invest in companies pursuing a strategy they favor (such as pursuing renewable energy opportunities), rather than pressuring one to move away from a core business in which it has long been involved.

Divest Harvard had been told that the endowment prefers to change companies from the inside rather than divest. But Harvard declined to push Chevron to shift its strategy and instead argued that it's better for shareholders to invest in other companies more aligned with their values.

Isn't that exactly what Divest Harvard has been saying?

The divest movement forces institutions to recognize a fundamental truth: there is no value-neutral way to invest. If an institution wants to invest according it its values, should it engage or divest? That depends on its theory of change. It also depends on how much trust there is between the institution and its stakeholders. Without trust, engagement will always sound like a remedy too little, too late against an existential crisis.

WORTH FIGHTING FOR

One Sunday night in the fall of 2019, Divest Harvard prepared for a protest at the annual Harvard-Yale football game. The meeting was in the ornate common room of Adams House, a dormitory just off Harvard Yard—a scene of oriental rugs and reusable water bottles; marble fireplaces and MacBooks covered in stickers; heavy drapes and fun-size candy wrappers. Students sat cross-legged on upholstered couches and plastic folding chairs. One placed a stack of library books on the floor, among them Uri Ben-Eliezer's *War over Peace* and *The War of Desperation*, by John Laffin.

They spoke passionately about systemic change and institutional intransigence, but they're still students. One recounted talking to a *Financial Times* reporter who was one of the most knowledgeable people he'd ever met on divestment. It was a surprise. "I've never read any of her stuff," he said, "because it's behind a paywall."

The meeting opened, as it always does, with a student sharing a story about environmental justice—an "EJ story." This week, it was about the wildfires in California. Discussion shifted to the business at hand: protesting the Harvard-Yale football game the Saturday before Thanksgiving break. Divest Harvard had a four-hour "nonviolent action training" scheduled for later that week.

It had been coordinating with Fossil Free Yale, whose members claimed they could get hundreds of people involved for a halftime protest. The Harvard students were skeptical. "A lot of people are going to be drinking," one student said. "It will be hard to get them to engage in activism."

Ilana Cohen pushed the group for ideas. "Something really serious is happening that both these universities are implicated in. We can't go with business as usual and just watch a football game." There were ideas for a sit-in. Or a funeral procession for planet

Earth. Or sneaking smoke bombs into the stadium in the marching band's instruments.

One student cautioned, "It will be easy for us to get villainized." But Cohen was undeterred. Being disruptive was the point. "This is an emergency," she said. "This is a crisis."

Sure enough, a month later Divest Harvard and Fossil Free Yale organizers took the field during halftime. They unfurled banners and chanted at the fifty-yard line as police officers in neon vests started walking toward them. Both football teams continued warming up. The cheerleaders continued their routines on the sidelines.

As the police approached, the protestors sat down. They weren't leaving. As they sat, a single student from the stands jumped the wall and sprinted toward them. He joined the protestors and sat. Another did the same. Then another. Then two more, their jackets flapping behind them as they ran. Ten minutes into the protest, six friends jumped the wall and a dozen more came from the other side. Suddenly dozens and dozens of students and alumni—too many to count—crowded the stadium stairs, jumped over walls, and stormed the field in Yale blue and Harvard crimson. The protestors were no longer sitting but standing, cheering, and jumping together with arms in the air.

The protestors then sat, the crowd now covering nearly forty yards from sideline to sideline. The game, televised nationally, was delayed by almost an hour. For the first time in its history, ESPN SportsCenter used the phrase "climate change." News of the protest was picked up by the *New York Times* and a dozen other papers that afternoon. Presidential candidates and climate activists promoted the protest and tweeted their support.

The fight over divestment is ongoing. After the game, the Harvard administration released a statement reiterating its position, making it clear that their minds were unchanged.

Yet, at least for one November afternoon, Divest Harvard had grabbed the nation's attention. "That moment, when we saw people running onto the field was just really incredible," said Schwartz. "I saw organizers around me crying because it was such a beautiful moment."

6

WE'RE GOING TO NEED A BIGGER BOAT

THE ALMOST ASCENDANCE OF IMPACT INVESTING

Despite the friendly audience, Jed Emerson was not looking forward to his speech. No crowd likes to be told that it's failing, especially by one of its heroes.

It was October 2017 at the Social Capital Markets (SOCAP) conference in San Francisco. This is where impact investors and socially minded entrepreneurs gather annually to compare notes on their efforts to foster a more just and sustainable economy. More than twenty thousand of Emerson's fellow travelers have made the pilgrimage since SOCAP's founding in 2008.

Emerson was a natural choice to give the keynote. He's worked as an activist for decades, toiling away in areas where our economy has failed people or the environment. After graduating from college in the 1980s, Emerson ran a nonprofit for runaway youths. Always something of a rebel, he was described as a "rail-thin, twenty-six-year-old chain smoker with a signature black leather jacket, jeans, boots, and no car." Over time, he grew frustrated with philanthropy. Too often, good intentions didn't translate into real change.

So he set out to find a better way. In the 1990s, he helped establish the field of venture philanthropy, based on the idea that capital markets can help solve social problems, and don't just cause them. Before then, a philanthropist tended to be either a capitalist or a critic. With creativity and tenacity, he fused the two. In the process, he helped lay the foundation of what would become impact investing. Impact investors explicitly seek social or environmental impact alongside financial return. They do so by funding mission-oriented companies or helping ordinary companies become more responsible. Today, there are 1,300 organizations managing $500 billion in impact investments worldwide.

The SOCAP conference is built in Emerson's image: part serious investor convention, part Grateful Dead powwow. Emerson once described it as "Burning Man meets Wall Street." It can feel self-congratulatory, an annual party for attendees to celebrate themselves. Emerson's speech was many things—critical, clear-eyed, a call to action—but self-congratulatory it was not.

"Sometimes he's angry when he says this message, because he's been doing this for a long, long time," the host warned. "And I'm excited for you to hear—because this will be the first time that I hear it as well—the latest from Jed Emerson. Please welcome Jed." The crowd eagerly applauded. Emerson emerged from backstage and removed a hoodie, revealing a black T-shirt with a clenched fist under the word RESIST. People clapped louder. A few whooped.

As Emerson settled behind the podium, the crowd quieted. "We, as a community, have grown lazy," he told the audience. His eyes scanned the auditorium. "We are at risk of taking the lowest path to the nearest point."

Emerson believes that capitalism is failing our country and that the tools we are relying on to fix it will not be enough. We are settling for the superficial while leaving the fundamental untouched. *We are at risk of taking the lowest path to the nearest point.*

Even today, impact investing represents well under 1 percent of global equity value. The field has generated nearly as many foundation-funded white papers as it has tangible victories in the fight to transform capitalism. Skeptics see pseudo-philanthropic investors piling into a narrow set of social enterprises, including solar panel manufacturers in sub-Saharan Africa and women's scarf-making collectives in India. Nothing against them, but they aren't quite transforming the Fortune 500.

We see something different. We see impact investing as the lab in which capitalists work to reform capitalism. We think that impact investors are developing a better way to build companies by focusing on more than just profit. The best impact investors are using the tools of active and engaged ownership to transform companies around a deeper purpose. These tools are applicable to every company in every sector. Ultimately, we think all investors should adopt this approach—that all investing should be impact investing.

But broader adoption is being held back by two debates that plague the field: First, how do you measure social and environmental impact? Second, do you have to give up financial return in order to have an impact? Traditional investors see themselves as maximizing a single variable: financial return. Now impact investors are trying to maximize at least two. Doesn't that require a trade-off?

In this chapter, we'll see how impact investing straddles the worlds of philanthropy and private equity. We'll see how these origins led to the two debates above and how those debates reflect most investors' confused understanding of what really builds long-term financial value. Finally, we'll see how a few investors in the public markets are beginning to apply the lessons of impact investing. They show what's possible if all investing becomes impact investing.

Emerson argued that the efforts being feted at SOCAP "require no real, profound critique of current practices within financial capitalism." He told the audience, "Nor do they require real change in our own behavior aside from adding a few funds to our portfolios here or augmenting a reporting process there." Rather than taking on the fundamental challenges of capitalism, Emerson worried, impact investors just want to think a little better of themselves and sleep a little more soundly at night. His speech was a call for them—and for Emerson himself—to aim higher.

Though he is right to worry, we think there is reason for hope. But for impact investors to reach their highest potential, they must first revisit their history.

STRANGE BEDFELLOWS

In 2007, a year before the first SOCAP conference, a group of investors and philanthropists sat around a conference room table at the Villa Serbelloni, in Italy, overlooking Lake Como. Surrounded by fifty-three acres of parkland, the villa is a natural home for big ideas. Indeed, that's what it's for. An American, the philanthropist and art connoisseur Principessa della Torre e Tasso (Ella Walker, granddaughter of the creator of Canadian Club whiskey), gave the villa to the Rockefeller Foundation in 1959 "for purposes connected with the promotion of international understanding."

For the past fifty years the villa has housed the foundation's Bellagio Center. It has used the space to discuss everything from the future of agriculture in the developing world to increasing access to vaccines and HIV/AIDS treatments. In 2007, the foundation convened a meeting around a similarly ambitious goal: How can the private sector be enlisted to help solve the world's problems?

The United Nations estimates that it will cost between $5 trillion and $7 trillion to eliminate poverty, protect the planet, and ensure that all people enjoy peace and prosperity by 2030. Whatever one thinks of the accuracy of such a broad figure, one thing is certain:

fixing the world's problems will be expensive. And it's not clear where the money is going to come from.

This fact wasn't lost on Antony Bugg-Levine, the Rockefeller managing director who had convened the conference. So he came up with an idea: What if the private sector could be harnessed to help meet these challenges? After all, the Fortune 500 alone has an equity value of $23 trillion. Within that context, the United Nations' cost estimate seems within reach. It was here at the Bellagio Center that Bugg-Levine coined the term "impact investing."

The idea of using business for good is not new. Early impact investors drew encouragement from social entrepreneurs such as Muhammad Yunus, the founder of Grameen Bank in Bangladesh. After the country's 1974 famine, Yunus provided small loans to impoverished entrepreneurs. Grameen Bank now serves more than 100,000 borrowers, and in 2006, Yunus was awarded the Nobel Peace Prize for his work.

More recently, the field has been led by pioneers such as Nancy Pfund of DBL Partners. After starting her career in traditional finance, Pfund launched an impact fund focused on small companies with big social potential. Pfund has led DBL to success through early investments in clean-tech companies such as Tesla and SolarCity. Starting with a $70 million first fund, the firm has now extended to a third fund of $400 million. In addition to its social and environmental impact, investors in DBL's funds have quadrupled their money, according to cofounder Mike Dorsey.

In the last few years, traditional private equity firms have jumped in—first Bain Capital and TPG and more recently KKR, Apollo, and others. When Massachusetts governor Deval Patrick left office in 2015 to found Bain Capital Double Impact, he set out to further the idea that business can be a force for good. "I feel like I've spent a lot of my career confronted by what turns out to be false choices," he said, "and in many cases this notion that you have to trade return for impact—that you can't do well and do good—is an example of one of those false choices." This is the fund that we helped build in our time at Bain Capital.

Bugg-Levine and Emerson wrote a book together that described the objectives of impact investing in almost spiritual terms: "At the conclusion of our life, each of us can look back on the path we took and say . . . we worked together, building on our relative strengths to create a greater, more sustainable whole." This theme of building on relative strengths resonates in a field that from its very beginnings has drawn on two very different traditions.

The first is philanthropy. For many, impact investing is just a new tool in the philanthropist's tool kit. In Bangladesh, Yunus could have donated money to reduce poverty. He could have built new schools. He could have founded an advocacy group to influence government policy. Instead, he started a bank. He did so on the principle that loans could do more good than charity. Compared to other forms of philanthropy, impact investing has the benefit of being self-sustaining. If the investment is successful, its revenue should offset its costs, and any profits can be reinvested in the mission. It also encourages discipline in recipients, because impact investors expect to be paid back.

The second tradition is conventional investing in the mold of private equity. Broadly, equity investors come in two stripes: those who invest in publicly traded companies and those who invest in private ones. The latter—known as private equity investors—can't buy and sell the shares they own on the stock market. Consequently, they usually hold their investments for years. Many private equity investments are sizable enough to influence or control the company. They have developed sophisticated corporate governance models and ways to help with strategy—things that can be very useful in an active-owner model of impact investing.

Philanthropy and private equity: strange bedfellows, indeed.

With these parents, it's not surprising that the children, impact investors, are confused. Those who lean toward philanthropy put social impact ahead of financial returns. They are called *concessionary investors* because they are willing to concede some financial return in order to have more impact. They include philanthropists such as eBay founder Pierre Omidyar and Bill Gates.

Those who lean toward private equity are called *market-rate investors.* They see impact investing as an attractive strategy for making money and doing good at the same time. This is what Bain Capital Double Impact and similar funds try to do. They believe that if they focus on impact, their investments will perform at least as well as, if not better than, investments that ignore it.

Unfortunately, a lot of impact investing falls into the gray area in between. If every impact investor wore a hat that put him or her into one camp or the other, these conflicting origins might be manageable, but in practice it's often not clear which side of the fence an investor is on. This confusion leads many people to suspect that *all* impact investing is concessionary in some way. Many new funds reject the label "impact investing" because they worry that it will always be perceived as less serious—incapable of earning a competitive financial return.

And even when impact investors do demonstrate high returns, many skeptics still aren't satisfied. They start with the premise that you can't do two things at once, so they assume that impact investors with high returns must be faking the impact—a practice called "greenwashing" in industry parlance. As an impact investor, you are usually seen as either a sheep among wolves or a wolf in sheep's clothing.

WITH OR WITHOUT YOU

Into this burgeoning field burst U2 front man and philanthropist Bono. In December 2016, the $100 billion private equity firm TPG announced that it had joined forces with Bono to launch the Rise Fund, the world's soon-to-be-largest impact investing fund. What's more, TPG announced that it had solved one of the field's most vexing problems: how to measure impact. Two years after the fund's launch, a *Barron's* headline captured the excitement: "TPG Cracks the Code for Impact Investing."

When Bono joined forces with the Rise Fund, he was outspoken in his criticism of the existing field of impact investors. He described the current state of impact investing as "a lot of bad deals done by good people." Some took umbrage, pointing out that Bono himself was probably a better singer than investor. But many at the time shared his assessment of impact investing. Larry Kramer, the head of the Hewlett Foundation and dean emeritus of Stanford Law School, believes that the field risks "just plunging ahead based on arguments that are imprecise or inaccurate." Asked how people have reacted to his critiques, Kramer remarked that he often feels like someone "who just entered a church and told the congregants that their God does not exist."

But TPG intended to help solve the measurement conundrum. And it would do so by borrowing a tool from private equity.

When private equity investors evaluate new investments, they usually project a multiple of money (MoM) return. MoM estimates the number of dollars a firm will get back for every dollar it invests. It doesn't matter if the dollar is made in pharmaceuticals or horse breeding, investors can use MoM to compare returns across deals. Together with the social impact advisory firm the Bridgespan Group, TPG announced that it could now do the same for impact measurement.

It called its new metric Impact Multiple of Money (IMM). It promised what had hitherto escaped most impact investors: a single quantitative measurement of impact to compare opportunities. IMM would translate whatever social impact an investment had into a single dollar figure, which TPG could then compare to the amount of money it had invested.

The process behind IMM is simple enough. For any potential investment, TPG finds research that quantifies the company's expected impact, translates that impact into a dollar value, and compares that value to the size of the investment. It adjusts that figure by how confident it is in the research. TPG would make an impact investment only if the IMM was over 2.5—that is, if it expected a return of more than $2.50 of impact for every $1 it invested.

For example, TPG invested in AlcoholEdu, an online training platform established to help curb alcohol abuse among college students. Based on one randomized controlled trial, TPG calculated that AlcoholEdu's programs would save thirty-six lives among the 2.2 million students who would use it over the following five years. TPG then assigned a $5.4 million value to each human life and—wait, *what?*

In its impact calculation, TPG used the US Department of Transportation's "Value of a Statistical Life." This figure is based on how much people are willing to pay to reduce the risk of death. For example, if people are willing to pay $1,000 to reduce their annual risk of death by 1 in 10,000—let's say by paying for a special car safety feature—then the Department of Transportation places a $10 million value on each life.

This calculation seems oddly precise. Indeed, even the Department of Transportation report TPG cited presented a range of values from $5.4 million to $13.4 million based on different studies. The Department of Agriculture uses yet another estimate—$8.9 million. What's the right number? This is only one area in which calculating the IMM requires finger-in-air assumptions and analytical gymnastics to create a single, seemingly simple output.

Financial models are useful tools, but they often reveal less about the world than about the people who build them. A model inevitably reflects the biases of its modeler. If a clever analyst wants a high IMM, he will find a way to produce it. Considering an investment in McDonald's? Food is necessary to live; there are plenty of reports available to support that. ExxonMobil? Oil heats houses, and gas fuels cars. You get the idea.

Before diving further into the rabbit hole of measurement, we might stop to ask: Why the focus on impact measurement in the first place? Can't impact investors just focus on investing in and building companies that have a clear, beneficial social purpose? Can't they focus on companies that sell only products that make customers healthier, pay workers fair wages, train them to take on higher-level work, and take the more sustainable option when it's economically viable? Can't they just get on board with the same sort of standardized ESG metrics being adopted in the public markets already?

This obsession reflects the field's philanthropic roots. The nonprofit world is swept up in the "effective philanthropy" movement, in which organizations are providing more rigor and data to demonstrate their own impact. Though this is driven largely by donors, the IRS has strict rules on who can register as tax exempt. When impact investors debate what should count as an impact investment, they are joining an old conversation about what counts as doing good. If you donate money, you want to see evidence that your money is not wasted. If your organization claims to be tax exempt, the IRS wants to make sure it deserves the status.

When we look at impact investing as a new tool for nonprofits, this focus on quantifying impact makes sense. But if instead we approach it as a better way to invest, this focus may be holding us back.

In TPG's view, IMM creates for impact what MoM does for financial return: a single, comparable figure to optimize. But perhaps both metrics—IMM and MoM—exacerbate the myopic tendencies of fiduciary absolutism. We should remember the English philosopher G. K. Chesterton's admonishment from 1908:

> The real trouble with this world of ours is not that it is an unreasonable world, nor even that it is a reasonable one. The

> commonest kind of trouble is that it is nearly reasonable, but not quite. Life is not an illogicality; yet it is a trap for logicians. It looks just a little more mathematical and regular than it is; its exactitude is obvious, but its inexactitude is hidden; its wildness lies in wait.

We are not antimeasurement. To operate a business effectively, you have to have focused metrics to guide managers. We do want practical measurement. Humble measurement. Measurement in line with what we saw in chapter 4: simplified, standardized, audited, and mandatory for all companies.

Jed Emerson offered his own critique: "As we know from the history of economics and finance, a single metric can't reflect more nuanced aspects of value or impact. Simplified metrics tell us how we are thinking today but not what we truly seek to know."

The quest to find a single number that captures all social and environmental impacts is an old one. It's no different from the dream of utilitarian philosophers, who wanted to assess moral decisions based on the aggregate utility they created. Governments, philanthropists, and now impact investors have sought a single number representing well-being for centuries, all to little avail.

Impact investors should focus on transforming business models and industries, on leading a movement that eventually all investors can join. They should avoid endlessly arguing about what counts as impact or how to measure it perfectly. They should avoid debating how many angels can dance on the head of a pin.

CHALLENGING THE TRADE-OFF MENTALITY

Even the most experienced and cutthroat investors don't make money all the time. How are we to know when poor returns are due to prioritizing impact, making bad investments, or suffering bad luck? It'll be many years before generalizable data on impact investment funds are even available. And so the suspicion persists that you can't both do well and do good—at least not through impact investing, at least not as a general rule.

As we'll show, impact investing isn't a surefire way to make money. No investment strategy is. But the most common criticism of impact investing—that it aims to achieve two things instead of one and therefore must involve concession on one or both dimensions—is based on outdated economic orthodoxy.

This is the critique we hear from all our friends in traditional finance. They spent their careers maximizing a single goal: financial return. With impact investing, they say, we added a second goal: impact. We went from maximizing a single-variable equation to creating a dual-variable one. After impact measurement, this is the second big question for impact investors: How can we possibly not be giving something up in exchange?

We used to joke that impact investing would not exist in fifteen years. If returns were bad, it would gradually fade into obscurity as a niche financial solution to narrow philanthropic problems. If returns were good, everyone would adopt the impact approach and the distinction between it and regular investing would become meaningless. The jury's still out.

Impact investing has been criticized in exactly the same way as all socially responsible investing and all efforts at corporate social responsibility—and for all the same reasons. But whereas we could previously show data that ESG performance was positively correlated with financial return, things become more complicated with impact investing. With impact investing, we've gone from single-variable optimization to trying to maximize two different things. Investors must—*must*—give something up in order to do good, the logic says. Otherwise, traditional investors would already be targeting social impact in their quest to maximize profit.

We've noticed that almost everyone who invokes the trade-off argument is talking about a trade-off with profit—that is, profit as it shows up on the income statement each year. But approaching the discussion this way stacks the deck, because all investments—whether in R&D or marketing or in a more environmentally friendly supply chain—cost cash today. The harder question is how to measure the benefit.

For that, we have to wade a bit deeper into a key financial concept: the valuation multiple.

For the uninitiated, consider this your crash course in corporate valuation. We've spent our careers obsessing over it, but we've boiled it down to its essentials. Let's take a company everyone knows: Coca-Cola. Is it a good idea to buy a share of Coca-Cola today?

At the beginning of 2019, the share price of Coca-Cola was $45. Ownership of Coca-Cola is divided into 4.3 billion shares, so the total equity value of the company was $196 billion. If you wanted to buy every share of Coca-Cola today and take control of the company, that's about how much you would need. So far, so good.

Is this a high or a low valuation for all of Coca-Cola? In accounting class, you learn a technical answer: you project all the future cash flow that Coca-Cola will ever generate and then discount that cash back to the present according to both how far in the future it is and how risky it is. If this sounds hard, you're right. We can't reliably predict next quarter's earnings, much less cash flows decades in the future. Investors, like all people, avoid doing hard things. Luckily for them, there's a shortcut.

Let's compare the share price to the annual earnings that Coca-Cola generated the year before. Coca-Cola's share price at the beginning of 2019 was twenty-two times the previous year's profit per share. This is called a *valuation multiple*, and—here's a little secret about the financial world—it's how almost all investors think about valuation. Rather than tediously calculating the company's future growth and risk, they just say that a share of Coca-Cola is worth twenty-two times its earnings.

Is that high, low, or just right? Let's compare it with PepsiCo.

PepsiCo's valuation multiple at that time was nineteen times its earnings. If you knew nothing else about these companies, you'd say that it would be cheaper to get a dollar of earnings by buying PepsiCo than by buying Coca-Cola. PepsiCo looks like a good deal, all else being equal.

Of course, there's the rub: "all else being equal." In reality, PepsiCo and Coca-Cola are very different companies, with different growth rates and risk expectations. Until we make a judgment on those, we just can't say whether Coca-Cola or PepsiCo is a better investment.

Congratulations, you passed Intro to Corporate Valuation.

Now let's try a valuation experiment: What would happen if Coca-Cola reduced the amount of sugar in its sodas to improve the health of its consumers?

It could lose profits if its customers were to eschew the new recipe. But let's say younger consumers prefer healthier products and prefer buying from corporations that care about their health. As these consumers age and become a larger part of the market, Coke's decision to lower the sugar content in sodas could be a significant draw. This could improve Coca-Cola's long-term growth. Further, let's suppose that the government creates a sugar tax in the future. In that case, Coke's reduction in sugar would also reduce its exposure to this risk. Both higher future growth and lower future risk should improve Coca-Cola's valuation multiple.

So is there a trade-off for Coca-Cola between providing a healthier product and creating long-term value for shareholders? Is it a good idea or a bad idea for Coca-Cola to cut the sugar content of the drinks it produces? Though the short-term impact on earnings is more tangible, the long-term benefits to the valuation multiple are immeasurable—not necessarily immeasurably large, just immeasurable.

(One of your authors was briefly a junior equity analyst at an investment bank, covering radio companies. At one point, he highlighted the risk that new technology could make radio obsolete in the long term. He was told, "Our investors only invest for three to six months at a time. They couldn't care less about that.")

Consider the even simpler case of maximizing profit just one weekend at a neighborhood bar. You could decrease the prices of drinks to bring in patrons. Would that increase or decrease profits? It depends. What about hiring an extra bartender? What about taking out an ad on Facebook? What about repainting the wall or buying a jukebox? It depends, it depends, it depends. Academic finance exists in the world of "all else being equal." All management—real management—exists in the world of "it depends."

And this is just when trading off short-term cash costs with short-term cash benefits. Though current profit is easier to measure, most of a corporation's value is not in today's profit. It's in the valuation multiple. Indeed, the present value of a company's existing contracts usually accounts for less than 5 percent of the company's value. For a typical company with a valuation multiple of fifteen, two-thirds of its value is based on earnings at least five years in the future. Once we start comparing short-term cash costs to changes in long-term growth and risk, it's even harder. At least we can wrap our minds around growth. But risk? Academic finance usually defines it as the day-to-day volatility of a corporation's share price, which surely feels like the measure of *something*—but not "risk" as most humans would define it.

Managers and investors want to maximize the value of their corporations. Most of this value is represented by the valuation multiple, which is determined by a corporation's long-term growth and risk. So why all the focus on current profits? It's like the joke about the drunk and the streetlight. A police officer comes upon a drunk late at night. The drunk is feeling around the sidewalk on his hands and knees below a streetlight. "What are you doing?" asks the officer. The drunk says, "I'm looking for my keys." The officer starts to help in the search. "Are you sure this is where you dropped them?" he asks. The drunk thinks. "No, I probably dropped them in the park." Officer: "Then why are you looking here?" Drunk: "There's better light."

We started this section with the challenge all impact investors are hit with: Don't you have to give up return in order to have impact? Isn't there a trade-off? Here's the answer: Yes. Of course there's a trade-off. All investments require a trade-off between a certain, measurable expense today and an uncertain, immeasurable return in the future. The question should be whether that trade-off will be worth it—whether running a company to maximize both impact and return could end up creating *more* return than running a company to maximize return alone. The question is whether, over the long run, financial return is a goal best achieved when you don't aim at it directly and obsessively.

BALANCE AND THE CORPORATE OBJECTIVE FUNCTION

Corporations have both a social service and a profit-making function, Professor E. Merrick Dodd, Jr., argued in *Harvard Law Review.* Though profit is important, we allow businesses to exist only because they serve the community. Dodd argued that corporations should adopt "a sense of social responsibility toward employees, consumers, and the general public." These are sentiments we hear frequently today, but Dodd was writing in 1932.

Dodd's article was a response to his academic adversary Adolf A. Berle, Jr. Half a century before Friedman, Jensen, or Meckling, Berle argued that all power in a corporation is granted by the shareholders and should be used only for their benefit. "Either you have a system based on individual ownership of property or you do not." This has been a long debate: What is a corporation's objective function—to benefit all stakeholders or the shareholders alone?

For a time, the debate seemed settled in Dodd's favor—not theoretically but based on how corporations were actually run in the mid-twentieth century. Managers saw it as their mandate to balance the interests of all stakeholders, and they believed that shareholders would be well served in the process. This was a world before fiduciary absolutism, and it looked a lot like the world we are pushing into today. Though impact investing is touted by many as a new thing, in many ways it is a return to the way corporations were run before fiduciary absolutism took hold, to a world in which managers were expected to balance various stakeholder demands and to consider shareholder returns as the prize for doing that well rather than an end in itself.

The very term "corporate social responsibility" can be traced back to Howard Bowen's seminal *Social Responsibilities of the Businessman* (1953). Bowen argued that profit maximization as the sole criterion of business success was fading into oblivion. That was not a radical view. At the time, 93.5 percent of executives polled by *Fortune* magazine agreed that business had a social responsibility beyond earning a profit.

Clarence Francis, chairman of the board of General Foods, voiced the sentiment even earlier, in 1948: "Today, most managements . . . operate as trustees in recognition of the claims of employees, investors, consumers, and government. The task is to keep these forces in balance and to see that each gets a fair share of industry's rewards."

He didn't seem particularly worried about maximizing a single variable. Instead, he invoked the idea of balance. This idea was best summarized in a *Harvard Business Review* article in 1960 entitled "The Trouble with Profit Maximization." Looking at how businesses were run, the author observed that they were "not consistent with *any* version of maximization. They are instead related to such notions as 'balance,' 'equity,' or 'adequacy.' The calculus of maximization will not fit these notions any more than it will define what constitutes an excellent dinner, a beautiful woman, a healthy man, a sound tax policy, or an adequate military establishment."

To criticize balance over maximization in business, he went on, is like "criticizing . . . physicists for their acceptance of Heisenberg's uncertainty principle. In both cases, the resulting body of theory is less precise, but it is also more realistic."

Some business operators—those on the front lines of capitalism—now echo this view. Here's Apple CEO Tim Cook: "If you want me to do things only for ROI reasons"—that is, a precise financial return on investment—"you should get out of this stock." He added separately, "When we work on making our devices accessible by the blind, I don't consider the bloody ROI."

Even the late Jack Welch, the legendarily tough former leader of GE, once said, "On the face of it, shareholder value is the dumbest idea in the world. Shareholder value is a result, not a strategy."

There is no clearer, more objective metric of success than winning in professional sports. Just check the scoreboard. Does that mean that coaches just go out and *maximize points*? Bill Walsh, a former head coach of the San Francisco 49ers and winner of three Super Bowls, argued that you should never focus on the score but rather on culture, on empowering players, on rigorous practice, on strategy. He titled his book *The Score Takes Care of Itself.*

Every small-business owner knows this basic truth: it takes a village to build a valuable, sustainable company over the long term. It takes deep investment by all stakeholders. Employees invest more into a company when they're fairly treated and engaged in the company's mission. The same is true for customers, suppliers, communities, and everyone else a company touches. Genuinely and consistently serving stakeholders creates the trust and goodwill that sustain great companies.

Cynics argue that impact investing must be inherently concessionary because it goes from maximizing one variable to maximizing two. In reality, it goes from optimizing one complex, interrelated set of variables to optimizing a slightly different complex, interrelated set of variables—all under great uncertainty. The belief is that social and environmental impact are just as important as any financial metric on the income statement.

We've seen this firsthand. We invested in Penn Foster, which provides job training to nearly 200,000 adults each year, helping them improve their workplace relevance and income mobility. The company focuses on jobs requiring "middle skills"—more education than a high school diploma but less than a four-year degree. And it does so without requiring student loans or government funding.

For-profit education has been out of favor since President Barack Obama cracked down on bad actors during his administration. To many investors, Penn Foster was an outmoded career-focused correspondence school. To us, it was an opportunity to take on a major problem: helping those unable to earn a living wage develop the skills to compete in an economy that is increasingly offshoring or automating work.

Penn Foster's "purpose with a profit" ethos was central to CEO Frank Britt and his team. Our investment came with a shared commitment to modernize the platform and extend its capabilities into new job-training areas, as well as to better measure which programs drive the best career outcomes. The transformation has created one of the most comprehensive upskilling platforms in the market, reaching new students across the country.

We also invested in Impact Fitness, a chain of Planet Fitness franchises founded by the successful socially oriented entrepreneur Chris Klebba. Impact Fitness operates these low-cost gyms in underserved markets in rural Michigan and Indiana, where obesity and diabetes are prevalent. Klebba and his team were driven by a mission: to improve the health of their communities. While all gyms worry about whether customers will sign up, Impact Fitness focused on making sure that customers who signed up actually used the gym—and got healthier as a result.

Some people are surprised that we saw a Planet Fitness franchise as an impact investment. But for us, it was an opportunity to show how a strong social mission can transform everyday businesses and also lead to better financial outcomes. We worked with Klebba to embed the company's mission in its charter and communicate it throughout the organization. Reorienting the business around this mission led to better employee and customer retention. We were able to recruit better managers. All stakeholders were more committed to the company because the company itself was committed to more than its own bottom line. As this kind of approach proves successful, we hope it will influence the way all Planet Fitness franchises and other small businesses are run.

Earlier in the chapter, we called impact investing the lab in which capitalists work to reform capitalism. This doesn't mean that they must have a perfect measure of impact. It doesn't mean that they have to give up financial return in order to have an impact. It just means that they have to build great companies by focusing on social as well as financial outcomes.

There are other investors who have signed on to this mode of thinking. If impact investing risks being too narrow, the United Nations–supported Principles for Responsible Investment risk being too broad. There are six principles; their purpose is to help investors commit to considering the environmental and social impact of their decisions. Since the program's launch in 2006, $80 trillion worth of assets have signed on to the PRI. The PRI are quintessentially UN: broad, lofty, and more revered in Europe than in America

The value of assets managed is astronomical—so much so that you might start to wonder why it is we haven't already fixed whatever is wrong with capitalism. The problem is in part that the principles themselves set a low bar. They are easy to sign on to but have no accountability to do anything differently.

Excitingly, we are also seeing investors in the public markets go farther—not just in tracking ESG but in investing in stocks to transform companies.

GOING PUBLIC

In January 2018, Apple's board of directors received a letter about the iPhone. The letter was signed by the hedge fund JANA Partners and the California State Teachers' Retirement System (CalSTRS). It detailed how iPhone usage can harm young consumers. JANA and CalSTRS cited one study showing that teenagers who spend more than three hours a day on electronic devices were 35 percent more likely to have a risk factor for suicide. *Three hours.* Each day, the average American teen with a smartphone uses it for more than 4.5 hours.

"We believe there is a clear need for Apple to offer parents more choices and tools to help them ensure that young consumers are using your products in an optimal manner," the letter said. The writers recommended that Apple form an expert committee to address the issue, begin its own research, offer better tools and options for parents, educate users, and report on its progress.

JANA and CalSTRS didn't make their request as concerned bystanders; they did so as shareholders. Together, they owned $2 billion in Apple stock. They also had the support of Sting, because apparently all high-profile impact investing requires a European rock star's backing.

"The results were, frankly, better than we could have even hoped for," said Barry Rosenstein, JANA's founder. "Within twenty-four hours the company capitulated and announced they would make changes." Apple responded by highlighting what it's already done but also announcing new features it planned for the future. Google, with its own sense of the zeitgeist, soon began offering "digital well-being" features on the Android.

The campaign was part of a new impact investing effort at JANA, led by a partner at the fund, Charles Penner. Penner made the case for why reforms at Apple would serve not just children and parents but shareholders as well. "Making people feel safe and supported in the ecosystem is incredibly valuable," he said. He sees this sort of impact investing as a way to divert the course companies are on and make them think longer term.

Unlike the sort of engagement we saw in the last chapter, impact investors in the public markets invest with the explicit goal of transforming companies in some way. They aren't just doing their duty to vote their shares in ways that align with their values; they are putting companies' social and environmental impact on par with their financial return.

"Was Apple happy we showed up?" reflected Rosenstein. "I can't imagine they were. Otherwise, they would have been on top of this issue before." But he believed that the test case had proved the power of investors to change public corporations for the better. "Having the ability to shine a public spotlight onto these companies and come to them with a solution makes them hard pressed to ignore it." He thought for a second. "And frankly, if we're able to push Apple to make changes, there's no company that exists in the world that we couldn't."

Though JANA's activist campaign against Apple was the hedge fund industry's most public foray into impact investing, JANA is only one of several firms pushing a new way of thinking into the public market.

Jeff Ubben, a cofounder of the hedge fund ValueAct Capital, has had a long career forcing change at the United States' largest companies. He took control of Martha Stewart Living in 2003 after Stewart went to jail. He helped push Steve Ballmer out as CEO of Microsoft in 2013. His track record is hard to beat, with returns of 20 percent per annum, compared with less than 5 percent for the S&P 500. And he's now pushing corporations not just to be more financially successful but to be more sustainable and longer term oriented as well.

He believes that the best way to do this is to go straight to the companies that are causing the biggest social and environmental problems. "The ultimate goal is to make your business totally sustainable because you're fixing a social or environmental problem while you're making your product with purpose to the customer," he said. The companies that will be the most durable over the long term will be those that make their money serving society rather than hurting it. He pointed to his investment in the AES Corporation, which had been a mainly coal-based power company.

At the time of his investment, many shareholders had assumed that AES would rely on coal until it was regulated out of existence or no longer economical. Ubben saw shareholders basically telling the company, "Pay your dividends until you are gone. That's the role you play in my portfolio." But Ubben saw a bigger opportunity in transforming the company.

He made a major investment, got himself placed on the company's board of directors, and helped reposition the company away from coal and toward renewable energy, something management had already begun pushing. As he did so, other shareholders took notice. They no longer saw just a tired coal burner that would pay dividends until it ceased to exist. Now it had growth prospects. The stock price jumped by 50 percent in a year. Through transformations like this, Ubben hopes to prove to other investors that serving society and the environment over the long term makes for better investing and better business.

Ubben recognizes that solving our largest problems requires changing our largest corporations. He believes he has to go straight to the problems because that's where systemic solutions are going to come from. About investors who focus more on small, virtuous companies, he said, "I guess it's fine to do that, but your impact is so, so, so small. It's so small." He continued, "What worries me is that people default to that because it makes them feel as if they're doing something."

Both Ubben at ValueAct and Rosenstein at JANA have launched their new impact efforts separately from their core investment funds.

Ubben stepped down as chief investment officer of ValueAct to lead the new fund. They mirrored another public investor with a more collaborative approach: Clifton S. Robbins and his Blue Harbour Group. Rather than waging big public campaigns the way Rosenstein and Ubben do, Robbins partners with managers who are open and willing to change.

In 2016, he fully integrated ESG into Blue Harbour's investment process, putting a company's social and environmental impact on a par with its financial return. "I think this is as important as the balance sheet," he said. "I really believe this is a new paradigm for smart investing." When corporate managers are not focused enough on ESG, Robbins thinks twice about partnering with them. "It makes me wonder two things. One, how is this manager evaluating other risks in the business? And two, is this someone we really want to put our name and money behind?" Ultimately, he believes that focusing on social and environmental impact will reduce risk and improve returns in his portfolio.

But all of this is easier said than done. Robbins believes that holding his companies, his fund, and himself accountable to social and environmental impact is important. He tells CEOs up front that they will have to answer for both their ESG and financial metrics. He began issuing a forty-page report to his investors about the progress of each company. Though we've grown accustomed to this sort of reporting by our largest companies, it is extremely rare in the investment industry today.

Underlying each of these three investors' efforts is a belief in the power of improving corporations over the *long term.* "How do we get the long term back?" asks Ubben. His answer: use the concepts of sustainability and stewardship to push the financial markets beyond the mindset of shareholder primacy. When companies are run for short-term profits rather than long-term stakeholders, when our investors provide no oversight, analysis, or meaningful governance, we'll end up with a few bloated public companies—"All big pigs," in Ubben's words—without purpose or direction. In a world that ignores the environment and society, Ubben said, "You're going to get shitty public companies and a shitty public market."

Robbins was more measured. Does he see public markets moving toward greater integration of social and environmental impact? "If you had asked me that question in 2015 or 2016, I would have said very few public market investors cared," he said. But in 2019? "Now almost everyone cares. We're getting to a point where ESG is table stakes." Soon, he believes, all CEOs will have to discuss social impact along with financial returns when they talk to their investors.

Rosenstein, Ubben, and Robbins are only three of the many public markets investors taking a new approach. "This has been a huge change just in the last twelve months," Robbins said. "I really think we're at a tipping point."

ALL INVESTING SHOULD BE IMPACT INVESTING

After the 2008 financial crisis, the British Parliament commissioned the economist John Kay to study how the United Kingdom's financial markets were helping or hurting the long-term performance of its companies. Perhaps the most clarifying point Kay made came in a press conference before he even began his work in 2012. As he summarized later, "The goals of equity markets are to operate and sustain high performing companies and to earn good returns for savers without undue risk." He continued, "Returns to beneficial owners, taken as a whole, can be enhanced only by improving the performance of the corporate sector as a whole." In the long run, the only thing that matters is how our equity markets help or hinder our corporations to become successful in this mission.

Too much of our thinking about finance is focused on the numberless intermediaries in our system. We lose sight of just how clear a task our financial markets have: to take the excess money from savers and turn it into valuable investments by corporations. We should judge our financial system based on how well it serves savers and the quality of the corporations it builds.

But today, two-thirds of Americans do not believe the financial system is benefiting the economy. Treating our capital markets like a casino can make money managers rich, but it doesn't make our economy prosperous. "When enterprise becomes the bubble on a whirlpool of speculation . . . the job is likely to be ill-done," wrote Keynes.

Today, impact investors in both the private and public markets are trying to change the way capitalism is done. They are following in the steps of Sir Ronald Cohen, whom many consider the father of impact investing. Since helping to bring venture capital to Europe in the 1980s and 1990s—and being knighted in the process—Cohen has led many of the first efforts in impact investing. He created one of the world's first impact funds, the first social impact bond, and the first social investment bank. He now chairs the Global Steering Group for Impact Investment. Gordon Brown, a former British prime minister, argued that Cohen's contributions deserved a Nobel Peace Prize.

Cohen believes that impact investing is the beginning of a revolution. Rather than considering only risk and return, all investors will consider risk, return, and impact. Soon, all investors will ask, "Did we really ever make investment decisions without considering impact?"

Ultimately, all investing should be impact investing. We should build a financial system that grows corporations for the long term in ways that have a positive impact on society and the environment. It's a revolution that government must support—especially in creating mandatory reporting standards that integrate impact metrics—but it's a revolution that cannot be led by government alone. "Governments are crucial," Cohen told an audience at Stanford in 2019. "If you don't get them as partners you'll achieve very little. But if you rely on them as partners, you'll achieve even less." Where we draw that line—exactly what role government should play in pushing companies to be good—is the subject of the next chapter.

Back at SOCAP, Emerson warned impact investors not to take the lowest path to the nearest point. Cohen echoed a similar sentiment. "There's a distinction between granular interventions and systemic change," he said. "It's one thing to get companies to begin measuring carbon emissions. That's not going to change a system." He concluded: "If you're thinking in terms of the fight to save capitalism, it must involve overthrowing the dictatorship of profit and putting impact at its side. That's the revolution."

7

DIVIDED AGAINST OURSELVES

GOVERNMENT AND THE LIMITS OF CONTROL

Not every Silicon Valley start-up can trace its lineage to the Book of Deuteronomy, but Ram Palaniappan's can. And he has raised nearly $200 million from a coterie of top-tier venture capitalists to show for it. Maybe other entrepreneurs should brush up on their religious studies.

"You shall pay them their wages daily before sunset," Moses preached to the Israelites, "because they are poor and their livelihood depends on them." That's the premise behind Earnin, Palaniappan's company.

Four out of five Americans live from paycheck to paycheck. Two in five can't come up with $400 in an emergency. Strapped for cash, they end up spending more than $50 billion each year in overdraft fees, on payday loans, and on interest in pawnshops. For the population that Earnin serves, one in eight Netflix transactions leads to an overdraft fee. That means that for every $100 Netflix makes off this population, banks make $35 in overdraft fees.

Against this, Palaniappan offers a new solution: "Our approach is that if someone has earned money, you should just give it to them." Earnin plugs into the time and attendance systems that major employers use. It knows on a daily basis how much employees have earned. If they want to cash out that money, Earnin will give it to them—up to $500 per pay period. When their paychecks eventually come, Earnin automatically deducts the same amount.

Rather than charging complicated fees or variable interest rates, Earnin asks users to pay an optional tip. The default is set to 10 percent, but users can lower it all the way down to zero. Based on one reporter's estimates, roughly 80 percent of users tip. Together, they voluntarily pay 3.4 percent of the cash they receive in a given month.

"This is one of the first pay-it-forward business models," Palaniappan said. "It's a community to help people pull each other up." Based on his own analysis, Earnin has saved its community $100 million each month in overdraft fees alone.

Earnin is trying to upend the payday lending industry and disrupt the standard industry practice of charging excessive overdraft fees and late fees on bills. With more payday storefronts in the United States than there are McDonald's, most such lenders charge astronomical interest rates on complicated financial products, snaring their customers in a cycle of debt that many cannot escape. The average interest rate on a payday loan is 400 percent, twenty-four times as high as the average credit card rate and one hundred times as high as the average auto loan rate. The average borrower takes out a loan for just $375 but ends up paying $520 in interest and fees because of repeat borrowing. Eighty percent of payday loans are rolled over, and three-quarters of all loans are to borrowers who have taken out eleven or more loans in the last year. They're trapped.

Earnin is one of several companies trying to serve these customers rather than impoverish them.

Another company, Elevate, charges interest rates one-quarter those of the competition and rewards good behavior by cutting interest rates further. It helps borrowers repair damaged credit scores by reporting good behavior to the credit bureaus. Elevate boasts that it has saved its customers $5.6 billion relative to payday loans. Its goal is to help its customers build self-reliance by ending their cycle of debt, providing financial education, and capping its own profit margins.

Governments have long tried to regulate the short-term lending industry. Indeed, if lending is an ancient profession, government efforts to control lenders are an ancient form of regulation. The Indian *Arthashastra* set a maximum rate of interest more than 2,300 years ago, and both Islamic law and the early Christian Church banned the charging of interest entirely.

Despite its good intentions, Earnin has now come under legal scrutiny. It's being investigated in eleven states and Puerto Rico for evading state usury laws. A $5 tip on $100 of cash repaid two weeks later is the equivalent of a 130 percent annualized interest rate. This is lower than that of most payday lenders but still above many states' legal limits.

We've spent much of this book describing how capitalists are trying to reform capitalism. But what about government? To many people, payday lending is representative of capitalism's inherently adversarial nature. There are malevolent corporations and the innocent masses, and the government's role is to fight the bad and protect the good. In his acceptance speech for his renomination for the presidency in June 1936, President Franklin Delano Roosevelt spoke of "small businessmen and merchants" protected, as early as the eighteenth century, from "economic royalists": "Against economic tyranny such as this, the American citizen could appeal only to the organized power of Government." Given all that is wrong with our economy today, why can't the government reform corporations to be good? Why can't it force responsibility by fiat?

Though the payday lending industry is a perennial target of public ire, recent calls for government intervention have focused on a new villain: share buybacks. Over the last decade, public companies have bought back $4 trillion of their own shares. To critics, this is money that could have—*should have*—been invested instead in workers and innovation, sustainability, and social responsibility. If we see our economy as inherently adversarial, then the situation is clear enough. Shareholders are cashing out while stakeholders suffer, and perhaps the government is required to balance the scales.

The same justification applies to legislating against excessive executive compensation. However, in all three cases—payday lending, share buybacks, and executive compensation—this adversarial view of the economy risks leading us astray. We risk focusing on politically salient symptoms without addressing the structural problems in our system. We risk trapping our government in a never-ending game of cat-and-mouse, where regulators chase the regulated around and around, always one step behind. Worse still, when we rely on government to dictate morality, we risk entrenching the very view that we should seek to uproot: that companies have no responsibility beyond the letter of the law; that businesspeople should always play "so close to the line that you get chalk dust on your cleats."

This is not to say we don't need "the organized power of Government" to deliver economic prosperity with equity and fairness, justice and responsibility. We do. We need structural reforms in our financial system that impose new duties on investors and new incentives for long-term stewardship. We need radical changes in how we price externalities (the costs of corporate activities that are borne by society, such as carbon emissions), provide public goods, and enforce fair bargaining and competition. We need better and clearer rules and more consistent referees. But above all else, we need to reconceptualize our economy as a shared project—essentially cooperative rather than adversarial. Ultimately, building companies that serve stakeholders over the long term is a goal shared by all of us.

As we've seen consistently throughout this book, most shareholders are stakeholders, too—in fact, they are stakeholders first and foremost. If share buybacks and executive compensation are excessive, most shareholders are ill served, just as the country is ill served. When we approach our economy in adversarial terms, we are dividing each of us against ourselves. We are unintentionally enforcing the view that what we do as economic actors should be held separate from and, often, in opposition to what we do as political actors. But each of us is one person, and together we are one people. In business and in government, we should act with integrity across all our roles. The problem is rarely malicious intent. More often it's a system that has dangerously skewed our shared long-term interests.

So long as we see our economy as inherently adversarial, the fighting may never end. After all, we are in our third millennium of unsuccessfully regulating payday lenders.

THE SCOURGE OF SHARE BUYBACKS

From 2008 through 2017, the S&P 500 returned $4 trillion to shareholders in the form of share buybacks, with another $1 trillion of shares bought back in 2018 alone. Together with dividends, that amounted to a stunning 88 percent of all profits. Much of that value came from the 2017 Tax Cuts and Jobs Act, which was meant to boost investment. In 2018, 60 percent of the money corporations saved from the tax cut was used to repurchase stock. To critics, this is fiduciary absolutism at its most self-destructive. Instead of investing that money for the long-term benefit of the corporation and its stakeholders, we are letting it go to enriching shareholders.

It wasn't always this way. Until 1982, share buybacks were essentially illegal, having been banned as stock price manipulation after the Great Depression. But then John Shad—the first Wall Street banker to lead the Securities and Exchange Commission in fifty years—changed the rules, unleashing what would become a torrent of share buybacks. As one commentator put it, our corporations are transitioning from a "retain-and-reinvest model to downsize-and-distribute."

What is a share buyback? Let's take Boeing as an example. In 2018, Boeing disbursed $13 billion of cash to shareholders. For most of the twentieth century, companies did so through dividend checks. Boeing would divide the $13 billion of cash by its 580 million outstanding shares and send every shareholder a check for $22 per share. Share buybacks are similar, except instead of using that $13 billion to send dividend checks, Boeing uses it to buy up as many of its own shares as it can afford, returning cash to shareholders who elect to sell. Then Boeing retires those shares. With Boeing's equity now divided into fewer shares, each remaining shareholder owns a larger percentage of the corporation. That is, with the ownership pie now cut into fewer slices, the relative size of each slice increases.

For sleepy shareholders, there is not much of a difference between share buybacks and dividends. With a dividend payment, Boeing shareholders get checks for $22 per share. With a share buyback, shareholders sell shares or see their relative stake appreciate by a commensurate amount. In either case, Boeing is distributing the same $13 billion of cash to shareholders—moving the cash from its balance sheet to theirs. So why does Boeing go through the extra effort? In part because share buybacks tend to increase a corporation's share price. The market perceives a share buyback as a positive signal for the corporation and bids up its share price accordingly. When the Securities and Exchange Commission studied the impact of 385 buybacks over 2017 and 2018, they found that the shares of corporations that announced a buyback outperformed others by 2.5 percentage points over the following thirty days. This is a significant bump over a short period of time. Of all the things executives might do to increase their share price—invest in a new project, expand into new markets, make a promising acquisition—few will lead to results as predictable and immediate as a share buyback.

On the one hand, the $1 trillion that public companies used to buy back shares in 2018 makes sense. That was in the tenth year of an economic boom. When the economy is doing well, we would expect profits to be up, and when profits are up, we would expect companies to return cash to their shareholders.

On the other hand, buybacks are worth a closer scrutiny. Cisco Systems spent $129 billion on share buybacks from 2002 to 2019, more than it spent on R&D. Are buybacks coming at the expense of innovation—there and throughout our economy? And we've seen excessive buybacks precipitate trouble before: Lehman Brothers bought back $5 billion of its own shares in the two years before it declared bankruptcy.

Under an adversarial view of capitalism, the problem looks like this: share buybacks benefit shareholders to the detriment of everyone else, and it's time for our government to stop the practice. On the left, Senators Chuck Schumer and Bernie Sanders proposed legislation that would limit buybacks. "At a time of huge income and wealth inequality," they wrote, "Americans should be outraged that these profitable corporations are laying off workers while spending billions of dollars to boost their stock's value to further enrich the wealthy few." Their bill would ban buybacks until a corporation completed a checklist of socially responsible actions: pay a $15 an hour minimum wage, provide seven days of paid sick leave, and provide employee benefits such as pensions and medical insurance.

On the right, Senator Marco Rubio argues that share buybacks "provide evidence of foregone investment in the economy as a whole." As evidence, he shows that although the private sector used to account for 35 percent of the country's spending on basic research in the 1950s, it fell to only 15 to 20 percent in the 2000s. He joined the chorus in decrying buybacks' pernicious role in corporations and proposed amending the tax code to address it.

Some politicians want to go beyond just prohibiting share buybacks. Senator Kamala Harris seeks to fine companies that don't pay women equally. Senator Elizabeth Warren wants the federal government to revoke the corporate charters of companies that misbehave. As part of their "A Better Deal" proposals, the Senate Democrats put forth the idea of a tax break for "Patriot Companies" that pay fair wages, provide medical insurance and retirement benefits, hire veterans and the disabled, and keep jobs in the United States rather than offshoring them. Companies that fit this description would be eligible for a $1,500 tax break per employee each year.

So: would it work? Presuming share buybacks are the scourge they're purported to be, would legislation be enough to make corporations reinvest their money instead?

We have reason to doubt that it would. Other countries have experimented with forcing corporations to do good through corporate social responsibility laws. India passed a law that required large corporations to donate 2 percent of their profits to charitable causes. What happened? Firms that spent less than 2 percent of profits increased their spending, but firms that spent more actually reduced theirs. The threshold not only became a new floor, it became a new ceiling as well. And that's only for companies that comply. Two years after the law was passed, over half of India's largest one hundred companies had found ways to avoid meeting the 2 percent requirement.

Here's the risk with government intervention: When government mandates socially responsible behavior, it undermines corporations' taking responsibility themselves. By dictating responsibility as law, it entrenches the view that there is no responsibility beyond the law. And in the process, it risks making things worse.

Our government's decades-long campaign to rein in executive compensation provides a case in point.

FIGHTING FIRE WITH GASOLINE

In 1996, the Walt Disney Company fired Michael Ovitz after only fourteen months as president of the company. Before briefly leading Disney, Ovitz had cofounded Creative Artists Agency, the most powerful talent agency at the time. He was brought to Disney by its head, Michael Eisner. They immediately butted heads.

One reporter remembered a particularly telling interview with both Eisner and Ovitz.

> Finally I settled down and began: "Michael . . ."
>
> "Which one?" they both asked.
>
> "Well, I've always thought of Michael Eisner as Michael and Michael Ovitz more as Mike," I said. "So let's do it that way."
>
> To my surprise, Ovitz immediately said, "I want to be Michael." But Eisner objected. He would be Michael, he said, and perhaps Ovitz could one day graduate from Mike to Michael. With that, Ovitz uttered a few words of reproach and hung up.

"It was an inauspicious beginning," the reporter reflected.

Ovitz's tenure would be ineffective and brief, as Eisner never relinquished control. Ovitz couldn't have been too distraught, though. His severance package was worth $140 million. Not bad for a year's work.

After that came to light, shareholders sued. The case snaked its way through the courts for years. To the eventual disappointment of shareholders, the courts ruled that the board had acted in good faith. The judge did note, though, that Ovitz "should not serve as a model for fellow executives." Clearly.

The absurdity of executive compensation persists today. Across the economy, CEOs' pay has increased by 940 percent since 1978, while typical worker compensation has increased by only 12 percent. The ratio of executives' pay to the pay of an average worker has skyrocketed from only 20 to 1 in 1965 to 121 to 1 when Disney fired Ovitz in 1995 to 278 to 1 today. If the minimum wage had kept up with the growth of CEO compensation since 1990, it would have reached $23 by the mid-2000s, versus the $7.25 that it is currently. When Bob Iger, Disney's current CEO, was paid more than $65 million in 2018, even Abigail Disney, a granddaughter of the company's cofounder, called it insane.

Government has tried three things to curb executive compensation, each of which was either ineffective or made things worse: mandatory disclosure, limits on base salaries, and giving shareholders a vote.

As far back as 1934, the SEC began requiring that public companies disclose their executives' compensation. The rationale was that increased transparency would shame boards and executives into maintaining reasonable pay levels. In 1978, the SEC went farther and began requiring executive perquisites and benefits to be reported as well.

What was the result? Once they began to be made public, perks significantly *increased* as executives learned what was common at other firms and began demanding more. In fact, the same thing happened to executive compensation in Canada after that country mandated pay disclosure in 1993. Rather than working as a moderating influence, disclosure actually led to *faster growth* in executive compensation than before the law was passed. Why? The Lake Wobegon effect.

The Lake Wobegon effect is named after the idyllic fictional radio town where "all the children are above average." Most people believe that they are special. When the members of a random audience were asked to rate themselves on whether they were average, above-average, or below-average drivers, 90 percent believed that they were above average.

So what happened when companies published the salaries of their CEOs? Well, CEOs all believe they are above average, and every board—whose job it is to hire the best possible CEO—wants to believe it has hired an above-average CEO. CEOs will never compare their salaries to those of their peer executives and say, "Bottom quartile? Sounds about right." And so every negotiation ends with a further increase in the average pay—up and up, right through the stratosphere, mesosphere, and exosphere.

When new SEC rules requiring disclosure of CEO pay ratios took effect in 2018, the result was similarly disappointing. Researchers could not find a single company in the S&P 500 that made changes to its compensation program based on comparison to the average. Nineteen out of twenty companies didn't even mention the ratio in their rationale for compensation levels.

What about limiting salaries and requiring that high compensation be linked to strong performance? In 1993, President Bill Clinton made good on a campaign promise to curb executive compensation by working with Congress to limit the tax deductibility of individual executives' pay to $1 million. But although the law helped constrain executives' base salary, it allowed unlimited performance-based compensation.

The results were striking. Before the 1970s, only one in six CEOs in the S&P 500 received performance-based compensation. By the mid-1990s, it was nearly one in two, and today it's the vast majority. In 2018, the average CEO pay in the S&P 500 was $14.2 million. Only 8 percent of that was salary. Over two-thirds was in stock, either as options or awards. Stock-based compensation is meant to align the incentives of managers with the interests of shareholders. Shareholders want stock prices to go up. Now managers do, too.

But if a core problem with capitalism is public companies' obsession with short-term stock price, stock-based compensation is surely a core cause. With most of CEOs' pay tied to share price, they'll work as hard as possible to push up the stock's value. Rather than compensating executives on the same basis on which we hire them—vision, judgment, teamwork, determination, and effective balancing of competing interests—we reward them for increasing the stock price alone.

Government's efforts to limit executive compensation unleashed a wave of compensation practices that forced managers to focus almost entirely on stock price. In trying to limit a symptom of an unequal society, the government exacerbated the disease of fiduciary absolutism.

Finally, what about giving shareholders a direct say? The Dodd-Frank Wall Street Reform and Consumer Protection Act of 2010 mandated that all public companies hold "say-on-pay" votes. At least once every three years, corporations must give shareholders a chance to vote on executive compensation. The legislation lacks teeth—the vote is nonbinding, merely advisory—but at least it gives shareholders a voice.

The result should not be surprising given what we've seen throughout this book. Though most shareholders are outraged by executive compensation, our financial system is so complex, distant, and intermediated that the signal is lost. The votes now amount to perfunctory approval. Taking 2016 as a sample year, only 1.6 percent of firms received a failing vote. Most companies receive 90 percent approval or higher.

Excessive executive compensation benefits exactly one group: executives. And though the harms of inequality are spread across society, executive compensation is a price literally paid by shareholders. In 2019 at Disney, shareholders had the chance to vote on Bob Iger's $35 million compensation package. Despite dissenting voices, they approved it. We march in the streets, rallying against inequality and the spoils of the 1 percent. Yet here, where we in our capacity as shareholders have a direct say, our voices are nowhere to be heard.

QUESTIONING THE PREMISE OF ADVERSARIAL CAPITALISM

Lost in some of the debate around share buybacks is what happens to the cash that is returned to shareholders. We saw how Boeing's decision to either issue a dividend or buy back its shares had the same net effect: it would move $13 billion of cash from its balance sheet to the bank accounts of its shareholders. In either case, the cash does not disappear. Shareholders must either reinvest it elsewhere or spend it. (Or they can leave it in their bank accounts, in which case the banks use it to make their own loans or investments.)

In an ideal economy, mature companies with high profits but poor prospects will distribute excess cash to investors, who will then reinvest that cash into young companies with little profit but much promise. We should want those dollars to be used in the most productive way. Far better that than trapping cash in bloated or wasteful companies. Why is it that we hear so many calls to ban share buybacks, but so few to ban dividends? In part, it's because the logic above is clearer with dividends than with share buybacks. Corporations paying dividends to their shareholders is somewhat analogous to homeowners paying interest to their banks. But share buybacks just *feel* like financial malfeasance.

So how do we make sense of the sheer size of share buybacks over the last decade? Shouldn't our corporations be investing that money instead? In 1982, the year share buybacks were reintroduced into the American economy, the United States spent 2.4 percent of GDP on R&D. By 2017, this had increased to 2.8 percent. In fact, *corporate* R&D spending has been increasing at an even faster rate than this suggests, because it's been offsetting a retreat in government funding. While the federal government spent 1.9 percent of GDP on R&D in 1964, that fell to just 0.6 percent of GDP in 2015, leaving corporations to pick up the slack.

The story with other investments such as new factories or physical equipment is the same. New private investment has averaged 17.4 percent of GDP since 1950. In 2019, it was 17.5 percent of GDP, which was in line with its trailing twenty-year average.

How can it be that corporations are able to distribute so much cash to shareholders in the form of share buybacks while also maintaining investment? Part of the answer is that this is what you'd expect when the economy is so strong. Corporations are not just earning high profits, but central banks including the Federal Reserve are keeping real interest rates at historic lows. With cash so plentiful and debt so cheap, there's just a lot of money sloshing around the system.

The other part of the answer is that the total cash returned to shareholders is not unprecedented. A report by Goldman Sachs found that since 1880, the S&P 500 returned 73 percent of its earnings to shareholders through either dividends or buybacks. In 2018, this same ratio was 88 percent of earnings—higher than average, but still lower than a quarter of all years since 1880. When critics cite the growth of share buybacks over time, they often fail to mention that over the same period dividends have been declining as corporations have shifted the way they distribute cash to shareholders.

Some individual corporations may be abusing share buybacks. They may be forsaking valuable investments in innovation, sustainability, or their workforces in order to reap the nearer-term, more predictable share price appreciation from announcing a buyback. A central premise of this book is that corporations are focused on short-term profits rather than these long-term investments in their stakeholders. And research indicates that after a corporation announces a share buyback, inside executives take advantage of the predictable price bump to sell their own shares.

But the macroeconomic data suggest that share buybacks are not the root of what's wrong in our economy. They are a politically salient bogeyman—an easy target because they are technical and unintuitive. But our problems predate share buybacks and would continue unabated even if buybacks were outlawed. Just as with executive compensation, limiting buybacks through legislation could have unintended consequences. It could trap cash in less productive companies, hurting growth across the economy. It could reduce the number of corporations who choose to go public, keeping a larger portion of our economy in the private markets where there is less transparency and accountability. Most likely, it would just push corporations back toward issuing dividends.

When we view the economy as adversarial, excessive share buybacks look like yet another case of shareholders profiting at stakeholders' expense. But we can see now that when buybacks come at the expense of long-term investments in innovation, sustainability, and the workforce, most shareholders lose.

Think again about Boeing. At the beginning of March 2019, a share of Boeing's stock traded as high as $440. Within months, it would lose a quarter of its value as the problems with the 737 MAX were revealed. By the summer, new plane deliveries fell by almost half. Would Boeing shareholders be better served by share buybacks and dividends or by greater investments in safety and quality, in research and development, in workers and goodwill? For its part, Boeing announced in 2019 that it might not buy back stock again for several years, canceling earlier plans. "We've also taken action to further sharpen our company's focus on product and services safety," said Dennis Muilenburg, Boeing's chief executive. We should hope. That would serve shareholders and stakeholders alike.

Following the recession, BlackRock's Larry Fink began speaking out against excessive share buybacks. "It concerns us that, in the wake of the financial crisis, many companies have shied away from investing in the future growth of their companies," he wrote in a letter to CEOs. "Too many companies have cut capital expenditure and even increased debt to boost dividends and increase share buybacks." No shareholder is served well by a corporate sector that bleeds itself dry trying to prop up lofty stock prices with buybacks rather than investing in its people, its equipment, its research, and its future.

It's not just executive compensation and share buybacks that serve the typical shareholder poorly. Consider corporate lobbying. Corporations often end up lobbying against the interest of the typical shareholder, who on balance is hurt by special favors and bent regulations. For example, of the $3.4 billion spent on lobbying in 2018, $280 million was by pharmaceutical companies. Much of it was spent defending high drug prices. As we saw in chapter 3, most shareholders hold hundreds or thousands of shares through their investment funds. Whatever short-term increase in profits that the pharma lobbying won for shareholders was likely offset by higher costs throughout the rest of the system, including in shareholders' own lives as consumers. Most shareholders are hurt by excessive or anticompetitive lobbying—even though the lobbying is supposedly done for their benefit. Indeed, the first federal campaign finance law was written in 1907 not just to protect citizens from undue corporate influence but also to protect the *shareholders* of those companies, whose money was being used in ways that ultimately undermined their interests.

With regard to executive compensation, share buybacks, and lobbying, we see that the true interests of most shareholders align with the broader public interest in having an economy that is prosperous and fair, sustainable and just, and built for the long-term benefit of all stakeholders.

And so the adversarial view of capitalism breaks down. There aren't innocent people in need of government protection from inherently evil corporations. Instead, there is a system that has subverted our shared interests by encouraging short-termism above all. The reforms we need are systemic changes to how our financial system works so that the interests of most shareholders—long term, stakeholder oriented, systemwide—are fully reflected in the way our companies are run.

Sure, government can try to force corporations to hew to a strict checklist of apparently virtuous actions. But if by "responsible capitalism" we mean shareholders, boards, managers, and employees actually taking responsibility for their corporations, government cannot conjure up that response through mandate or endless bureaucratic intervention. Indeed, the farther government encroaches, the more it entrenches the idea that corporations have no responsibilities beyond the law.

Instead, we think government should focus on setting the groundwork that will allow our corporations to compete on which of them can best serve society. There is so much that is in the enlightened self-interest of corporations to do. The fact that they don't represents a failure in governance more than a failure in government. Government should focus first on fixing the underlying structures of our financial system so that our corporations are run in a way that actually serves our shared interests.

BUILDING FINANCE AROUND STEWARDSHIP AND ACCOUNTABILITY

In 2004, the investor Richard Perry owned shares in King Pharmaceuticals. Perry thought it was about to be acquired by rival Mylan Laboratories, which would be very good for him. Unfortunately, Mylan shareholders weren't so sure that buying King would be very good for them. So Perry did something remarkable: he bought 9.9 percent of Mylan shares with the intent of pushing the acquisition through. His interests were not aligned with those of most Mylan shareholders, though. That was because when he bought 9.9 percent of Mylan, he simultaneously entered into a financial contract so that he would actually make money if the Mylan stock price *declined.* He had the power to vote the shares in favor of acquiring King, and if it ended up hurting Mylan, he would profit twice over.

This is our financial system at its worst—a casino entirely divorced from any sense of stewardship or accountability. Today, no legal duty binds shareholders to act in the best interests of the companies they invest in. This has to change.

The United Kingdom offers one path we might take. In July 2010, it introduced a Stewardship Code for investors that has since been adopted in one form or another in twenty other countries. The code sets a high benchmark for stewardship "as the responsible allocation, management and oversight of capital to create long-term value for clients and beneficiaries leading to sustainable benefits for the economy, the environment and society."

Updated in 2019, the code has the aim of "making investors' stewardship efforts more transparent, thereby making them more accountable." It asks investors to "systematically integrate stewardship and investment, including material environmental, social and governance issues, and climate change, to fulfill their responsibilities." It promotes active engagement with companies and a responsibility to ensure a well-functioning financial system.

"Signatories are now required to explain their organisation's purpose, investment beliefs, strategy and culture and how these enable them to practice stewardship," the code stipulates. "They are also expected to show how they are demonstrating this commitment through appropriate governance, resourcing and workforce incentives."

The UK Stewardship Code is still voluntary—a "comply or explain" regulation with no teeth. "If the code remains simply a driver of boilerplate reporting, serious consideration should be given to its abolition," wrote one critic. Nonetheless, we think it provides a helpful model of how we might make the duties between corporations and investors reciprocal—by insisting that investors who vote shares on corporate resolutions be duty bound to act in the long-term best interests of those corporations.

A Stewardship Code would be a great first step—but further steps are required to align disclosures, compensation, and taxes with the long-term public interest.

First: board members themselves should better represent the stakeholders they are responsible for, and they should be elected for three-year terms rather than annually. Longer cycles would give shareholders fewer board elections to focus on each year and the ability to assess each one more thoughtfully. Rather than our current system, which vastly favors incumbents, all nominees with significant support should have full access to the election machinery that is currently reserved for internal candidates.

Second: we should end quarterly earnings guidance and make ESG metrics a standardized part of 10-K reporting, along with fully transparent reporting on lobbying and political contributions in all their various forms ("hard," "soft," "dark," and the rest). Money managers should be compensated on the basis of their long-term, rolling performance rather than on the total amount of assets they've gathered or how they performed just this year or this quarter. Rather than paying executives with options, we should allow newly empowered boards and shareholders operating under a new Stewardship Code to set long-term compensation schemes that reward building sustained corporate value and delivering on a corporation's purpose. Much of this should be performance based, but performance should be based on much more than share price fluctuations, which can be gamed through financial engineering and also reward executives for the broader performance of the stock market.

And third: we should consider ways to use tax policy to reorient incentives in our financial system. Some people have called for changing the capital gains tax rate to encourage longer-term investments. Others have called for a change in the way we treat interest payments on debt and for an end to the carried interest loophole, which allows investment managers to pay a relatively low tax rate. Any changes in tax policy can have wide-ranging and sometimes unintended consequences. However, we think questions such as these should be assessed according to a simple standard: whether or not they encourage strong, sustainable corporations over the long term that serve our broader interests.

THE WORK WE DO TOGETHER

Even with the reforms listed above, government would still have important work to do to ensure a vibrant and fair economy. Our partner Governor Deval Patrick liked to recall the words of former representative Barney Frank: government is the name we give to the things we choose to do together. Even if every corporation were the Platonic ideal and every investor the perfect embodiment of stewardship, there are still things that are possible to do effectively only if we do them together.

Government has always played an active role in building the US economy. The steamboat required a government monopoly to develop, and the telegraph required a government grant. The government gave so much land to the railways in the decade after 1862 that, taken together, they would have made up the largest state in the United States after Alaska and Texas. The steel industry required decades of tariff protection during the Gilded Age. Government-backed home loans supported suburbanization, and federally insured bank deposits support the banking industry.

"Capitalism in America was not arm's-length ideology," wrote Bhu Srinivasan in his history of the US economy. "It was an endlessly calibrated balance between state subsidies, social programs, government contracts, regulation, free will, entrepreneurship, and free markets."

Make no mistake: in a democracy, all government regulation is self-imposed. We can collectively decide that it makes more sense to have uniform car safety standards than have each company come up with its own. We do so even if that means paying more for cars. We can collectively decide that it makes more sense to impose a price on carbon than have each company incentivized to pollute at society's expense. We can do so even if that means paying more for energy. We are both citizens and economic actors, and sometimes we must be citizens first.

One of the clearest cases for this sort of self-imposed regulation is to make sure that corporations can't off-load their costs onto the rest of society.

PRICE EXTERNALITIES

Through the 1980s, the sulfur dioxide emitted by coal-fired power plants caused acid rain, severely damaging lakes, forests, and buildings. This kind of unaccounted-for harm is called an "externality"—an external consequence that the power plants imposed on the rest of society. Though the damage was something that we all bore, the power plants did not have to pay anything for causing it or cleaning it up.

So we, through our government, decided to put a price on this externality. In the Clean Air Act Amendments of 1990, we created what's known as a cap-and-trade system, which in effect makes power plants pay for the sulfur dioxide they emit and creates incentives for facilities to reduce their emissions.

The program was a success. Even though electricity generation by coal-fired power plants increased by 25 percent over the following fifteen years, sulfur dioxide emissions fell by 36 percent. Those lower emissions cut the amount of acid rain in half, generating an estimated $122 billion in environmental and health benefits.

Today, the most important externality that remains unpriced is carbon. Without a price on carbon, we are essentially subsidizing its production and consumption. This is a difficult issue for politicians because most of us are carbon consumers, and setting the right price for carbon is difficult. But even a small price would be better than no price, because it would be a step toward greater accountability for corporations and consumers alike.

The idea of pricing externalities extends beyond pollution. For example, the progressive group Americans for Tax Fairness argued in 2014 that Walmart's low wages cost taxpayers more than $6 billion in federal subsidies. "Walmart pays its employees so little," its report stated, "that many of them rely on food stamps, health care and other taxpayer-funded programs."

That particular estimate has been challenged, but it raises an important question: Should a company that pays its employees less than a living wage be asked to help pay for the increased public services they require? The logic could be applied more broadly to cover not just wages but also layoffs that impose a cost on the rest of society. Critics argue that this could reduce the demand for domestic labor by increasing its cost relative to foreign labor. In some cases this might be right, which is why we have to consider the price of externalities in the context of our trade policy. When we take steps to fix an externality within our own borders, we shouldn't allow trade policy to undermine those steps. If a car manufactured in the United States embeds the cost of carbon emitted in the process, we should apply that same cost to cars made abroad. "Just as we should globalize the world's economy in order to spread prosperity as widely as possible," wrote former chief justice Strine, "so too must we globalize those policies that reflect our best ideals."

Until we use our government to put a price on externalities, our most responsible corporations will be forced to compete against their least scrupulous competitors.

PROVIDE PUBLIC GOODS

Without public investment in infrastructure, public education, or basic R&D, no corporation could operate effectively. For example, though companies fund 82 percent of all development spending in our economy, they fund only 27 percent of basic research. Basic research is the type of investment that, though it has no readily applicable market use, eventually powers the next generation of development. Robust basic research spending is a public good, provided predominantly by the federal government and our universities.

Unfortunately, federally funded R&D spending as a percentage of GDP has been cut by nearly three-quarters since 1960. In 1960, the United States accounted for 69 percent of global R&D spending. By 2016, that figure was only 28 percent, with China close behind.

One public good we often take for granted is the provision of consumer safety through programs such as the Food and Drug Administration. Rather than putting the onus on individual consumers to figure out which products are safe and which companies they can trust, we do so collectively. This same logic underlies the avoidance of fraud through clear accounting standards; as we argued earlier, we think it applies to ESG disclosures as well.

Another idea we think has merit is the increased use of social impact bonds in the provision of public services. Social impact bonds are contracts with nonprofits and for-profit corporations to deliver on specific social outcomes. As with the Stewardship Code, the idea was pioneered in the United Kingdom, and it has been used to effectively reduce recidivism rates in prisons, for instance. It is an interesting tool with potentially broad applicability for applying private enterprise ingenuity to social and environmental problems.

ENFORCE FAIR BARGAINING AND COMPETITION

When private entities compete, stakeholders are well served. But this does not happen on its own. Indeed, power tends to concentrate until competitors choke one another out, creating monopolies and abusive conglomerates that are against the best interests of the public at large. Strong and vigilant antitrust measures must undergird any well-functioning economy. Many people worry that our largest technology companies now mirror the trusts of the robber baron era.

But it's not enough just to ensure fair competition between corporations. The government should also enforce fair bargaining between stakeholders. Low-skill workers, for example, lack the ability to bargain fairly against billion-dollar employers. This is one of the justifications for establishing a minimum wage and mandatory benefits, supporting unions, and limiting overly restrictive occupational licenses and noncompete clauses.

These sorts of laws can also set the constraints within which corporations innovate. If labor is relatively more expensive, corporations must invest to make it more productive. In the United States before the Civil War, the economy of the North was powered by free labor, while that of the South was powered by slaves. So "the North invested in labor-saving devices, in both agriculture and industry," wrote the economist Alan Greenspan and the historian Adrian Wooldridge, while "the South invested in slaves."As a result, 93 percent of the important inventions in the United States between 1790 and 1860 were patented in free states. Well into the twentieth century, GDP per capita in the North was more than double that in the South.

Of course, this is another area where coordinating with our trading partners is important. If we raise the cost of local labor but leave the cost of foreign labor unchanged, we shouldn't be surprised when competition forces corporations to move those jobs offshore. We should make sure our labor standards don't hurt us in trade deals.

One final area for enforcing fair bargaining is when there is asymmetric information—when customers don't have all the information they need or can't reasonably be expected to know what they're buying. For example, Americans regularly fail even the most basic financial literacy tests. Nearly two in five Americans do not know what compound interest is. As one economist noted after the Great Recession, "An overwhelming majority of borrowers who were receiving adjustable rate loans did not understand that their mortgage payment was not fixed over the life of the loan." The entire premise behind the Consumer Financial Protection Bureau is that if we look at the world as it is rather than as we want it to be, we can't expect the average consumer to go toe to toe with Wall Street and come out ahead.

WHAT ABOUT PAYDAY LENDING?

Now, at last, we can return to payday lending. Payday lending is regulated mostly by the states, with seventeen US states prohibiting payday lending or capping interest rates. Few states have garnered as much praise for their particular approach as has Colorado. Its reforms in 2010 began by recognizing that some Coloradans do face short-term cash needs. When other states have tried to outlaw payday lending, online lenders set up rent-a-bank schemes to continue operating. Outright prohibition works about as well as Prohibition did, which is to say, not at all. Though it was illegal to produce, transport, or sell alcohol in the United States from 1920 to 1933, the proportion of US GDP spent on alcohol never budged.

So instead of prohibiting payday lending, Colorado set standard terms that served borrowers' needs rather than trapping them in debt. They are the terms that individual borrowers would negotiate if they had the power and knowledge to do so. After 2010, all payday loans could be repaid in equal installments over six months with no penalty for prepayment. Now nearly three out of four loans end up being repaid early. The average interest rate is 115 percent—but even that is lower than in any other state that has payday loan stores. It better reflects the riskiness of the borrower than abuse by the lender.

Three years after the reform, researchers were able to see the effects: borrowers spent 42 percent less on payday loans, saving $40 million in aggregate, and defaults fell by 23 percent. Half the payday lenders in the state closed within three years. But that all happened while the number of loans *increased.* As the Pew Charitable Trusts noted, "The law's transparent pricing, realistic loan durations, and lower price limits have produced a market in which lenders succeed if borrowers repay loans as scheduled."

And so we can justify the government's intervention in the payday lending industry as enforcement of fair bargaining between sophisticated financial lenders and unsophisticated, low-income borrowers in a bind. Colorado is a good example of government generating an outcome that approximates fair bargaining between buyers and sellers of payday loans.

What does this imply for Earnin? Earnin's pricing could hardly be more transparent: use the service, and pay what you like. When government lawyers compute the implied interest from a user's optional tip, they are creating complexity where there once was simplicity. Indeed, Palaniappan built the business on an optional tip because he believed that would remove any temptation to exploit his users. If we justify government intervention in payday lending because of the complexity and unfairness of payday loans, it's hard to justify the government intervening in Earnin, which couldn't be much simpler or more fair.

And if we go one layer deeper, we might ask why 12 million Americans resort to payday loans in the first place. "It's not like these customers don't know what's happening," said Palaniappan. "They know exactly what their bank balance is." It's not so much that they're ignorant as that the system works against them. Take housing: "They can't afford a thousand-dollar down payment, so they spend $200 a week on housing instead of $500 a month." According to the *New York Times*:

> What fed the industry's growth were not emergency expenses but the increasingly unstable incomes of the working poor. As their hourly wages fluctuated at the whims of

> workplace-optimization software, payday-loan customers . . . borrowed to pay their rent or electric bills.

Indeed, seven of ten borrowers use loans for basic expenses, not emergencies. Perhaps the best regulation would be one of the structural changes we propose above: a stronger minimum wage that is also predictable and consistent.

Maybe there's another systemic flaw that we need to address instead of just focusing on interest rates: our outdated payroll system. Currently, $1 trillion of earned income is held up in payroll departments at any given time, while bills and expenses come in day after day. It seems punitive to cash-strapped workers that they have to wait two weeks for wages they earned today. That, after all, has always been Earnin's larger vision: *You shall pay them their wages daily before sunset, because they are poor and their livelihood depends on them.*

THE BILLIONAIRE COP-OUT

Yet, however important reform is, it does not absolve us from changing our businesses and economic behavior as well. Just as responsible corporations are no substitute for effective government, effective government is no substitute for responsible corporations.

In recent years, some of capitalism's greatest beneficiaries have rushed to pontificate on its flaws. Some, such as Warren Buffett, focus on the tax code, imploring the government to tax them more, please. Others, such as JPMorgan CEO Jamie Dimon and hedge fund billionaire Ray Dalio, have focused on the ways in which the government is failing to create an economy that works for everyone.

We should fight to fix our government, but remember that there is so much we can do through corporations in the meantime. It's convenient for those with power to argue that change must only come from our most notoriously logjammed institution. Good laws can encourage good behavior, but they are no guarantee of it. If we care enough about an issue to cast votes with our ballots, we should care enough to vote with our wallets. If we care enough to change how we legislate our country, we should care enough to change how we operate our companies.

Walmart and the Walton family, who collectively own roughly half of Walmart's stock, are not always held up as paragons of social responsibility. However, in one particular case, they did what few others would.

Like clockwork, mass shootings in the United States are followed by a cycle of mourning, advocacy, and despair; public outrage followed by policy proposals followed by the shelving of those proposals. But by 2018, Walmart had had enough. "The status quo is unacceptable," Walmart CEO Doug McMillon said. And so the company didn't accept it. Realizing its immense power over US commerce, it stopped selling assault-style weapons and raised the age requirement for all gun purchases to twenty-one. "We take seriously our obligation," Walmart's website reads, "to be a responsible seller of firearms and go beyond Federal law by requiring customers to pass a background check before purchasing any firearm."

Ed Stack, who bought Dick's Sporting Goods from his father in 1984, went even farther. After a gunman killed seventeen people at a school in Parkland, Florida, he decided enough was enough. "Watching those kids, listening to those parents—it had a profound effect on me," he told the *New York Times*. "It was at that point I said, 'I just don't want to sell these guns, period.'" Dick's not only stopped selling all assault rifles and high-capacity magazines, it destroyed the assault rifles on its shelves rather than return them to the manufacturers.

Are Dick's Sporting Goods' or Walmart's actions a panacea for gun violence? Of course not. But those companies did something.

The Waltons did something. Ed Stack did something. They

didn't wait for a perfect solution. They didn't write an op-ed maligning the government for inaction, all the while selling every firearm they could right up to the thin line of the law. They recognized the immense power their companies had and acted according to their values.

Though we should never stop fighting to make government more effective, we should not allow its dysfunction to let us off the hook. In claiming that *only* government can enact change, some business leaders pass the buck on to government to solve problems that they helped create and over which they maintain much control.

Our political system is an imperfect arbiter of corporate behavior. We must recognize that both our corporations and our government are flawed institutions. We blame corporations for short-termism, but Vanguard estimated that short-termism in government and the resulting uncertainty have created a $261 billion drag on our economy. For example, even though a broad swath of the country agrees that we need to combat climate change, our current administration has rolled back environmental protections as corporations have lobbied against the broader, long-term interests of their shareholders. In an otherwise unremarkable press release in 2019, the Department of Energy began calling natural gas "molecules of U.S. freedom" and "freedom gas"—not quite the prelude to a carbon tax.

Whereas in 1960, 75 percent of Americans trusted the government to do what is right most of the time, today only 20 percent do. So it's no surprise that 64 percent of Americans say that CEOs should take the lead on change rather than waiting for government to impose it.

Against these things, we need to do what work we can, where we can. "Business has to pick up the mantle when government fails you," said Rose Marcario, the CEO of Patagonia. Ed Stack agrees: "If you have ideas about how to solve certain problems, I think it's your responsibility as business leaders to speak up."

Adversarial capitalism starts from the premise that corporations are inherently bad and government is inherently good. But these are all just social institutions we collectively set up to build our society.

We can create an economy filled with corporations organized around a deeper social purpose that hold themselves accountable to that purpose and build business models in which purpose and prosperity are aligned: income for employees, profit for shareholders, taxes for the common good.

Government can facilitate a financial system marked by stewardship and accountability. Government must do the work that we can do only together. But in capitalism, the corporation is the fundamental unit of analysis. If we want to create a better capitalism, we have to build better corporations.

8

THE $23 TRILLION SOLUTION

REPURPOSING OUR CORPORATIONS

Early one fall morning in 1982, twelve-year-old Mary Kellerman had a cold. She lived in Elk Grove Village, a suburb of Chicago. Her parents gave her a Tylenol to ease the symptoms. An otherwise healthy girl, she was dead by 7:00 a.m.

Later that day, twenty-seven-year-old Adam Janus of nearby Arlington Heights had chest pains. He took Tylenol and soon died of what those around him thought was a heart attack. His brother and sister-in-law rushed to his home. Both had headaches and other signs of stress, understandable under the circumstances. Both took Tylenol from the same bottle. Both were dead within a couple of days.

Three more strange deaths occurred in the area in the following weeks, and authorities saw the pattern. All of the incidents had been caused by Tylenol capsules somehow tampered with and laced with cyanide. We still don't know who did it or why. Seven lives were lost in all.

At the time, Tylenol was one of the most popular over-the-counter drugs in the United States, controlling more than one-third of the acetaminophen market. For its manufacturer, Johnson & Johnson (J&J), the crisis had huge implications.

Though J&J manufactured products across dozens of business lines, from shampoo to baby powder, Tylenol accounted for more than 15 percent of its profits. And as a well-known, trusted consumer brand, Tylenol was inextricably tied to the company's reputation.

"There were many people in the company who felt there was no possible way to save the brand, that it was the end of Tylenol," said James Burke, who was the company's CEO during the crisis. Most experts agreed. Tylenol was doomed and J&J was in deep trouble—no normal company survives a disaster like this unscathed. But J&J is no normal company. Its highly principled response set the standard for how companies should manage a crisis. As the *New York Times* later put it:

> The Tylenol episode remains a textbook illustration of effective crisis management, and yet its lessons are all too often forgotten by the corporate world. One after another in recent years, companies like Equifax, Wells Fargo, United Airlines, Facebook and Toyota have fumbled their way through scandals—self-inflicted wounds in the main. What they tend to have in common are responses regarded as too slow or too grudging. Or both.

J&J's response is not just some lesson in public relations; it's an example of how a corporation with a deep sense of purpose can rely on it to guide strategy through difficult times. As we'll see, J&J's culture was grounded in a mission that transcended stock price, shareholder value, and quarterly earnings. We would do well to learn from its example.

We've argued that capitalism in the United States today is broken. Through hewing to the tenets of fiduciary absolutism, corporations have lost any sense of purpose beyond maximizing profit. Our deeply intermediated financial system has led to a Milgram-Nixon Syndrome in the corporate world: like Nixon, shareholders claim that they are too far from the action to be responsible; like Milgram's study participants, managers claim they're just following orders. This hurts all stakeholders—shareholders included. Most shareholders have an interest in the long-term success of the entire economy, not just the quarterly profits that drive short-term share prices.

This leads to the Two-Buffett Paradox: corporations are being asked to do two contradictory things. Maximize profit at all costs, says Buffett the elder. But do good things for society despite the costs, says Buffett the younger. The result has been a set of solutions that are often too hypocritical, superficial, or marginal to solve the larger problem.

We offer three proposals meant as antidotes to fiduciary absolutism, the Milgram-Nixon Syndrome, and the Two-Buffett Paradox. We believe that these proposals can transform Fortune 500 companies into organizations that reflect our values, and that they will unleash the power of corporations—that's $23 trillion of equity value in the Fortune 500 alone—to build the society we desire.

To create corporations that reflect our values, we propose the following:

PROPOSAL 1: Recharter corporations around a deeper purpose.

PROPOSAL 2: Hold them accountable to that purpose.

PROPOSAL 3: Align their purpose with their prosperity.

J&J's response to the Tylenol crisis illustrates a critical part of this vision; it is a classic case study in purpose-driven accountability. The turmoil at the online craft marketplace Etsy offers a more recent and controversial illustration. After becoming a public company, Etsy fell into a crisis of its own, as both its deeper purpose and its ability to profitably grow were thrown into question. Josh Silverman's leadership as CEO through this period reveals the importance and pitfalls of each of these proposals.

J&J and Etsy are imperfect—just like every organization humans have ever formed. Yet they reveal what's possible when corporations are built around a deeper purpose, when they are held accountable to that purpose, and when they align their purpose with creating prosperity for all stakeholders.

But first, we have a Tylenol crisis to solve.

THE CREDO

J&J was always meant to be more than just a profit-making enterprise. In 1886, two brothers established the company, but it was a third sibling, Robert Wood Johnson, who made the largest mark.

Johnson served as chairman from 1932 to 1963. He crafted what he called "Our Credo"—*credo* being Latin for "I believe"—in 1943, just before J&J became a publicly traded company. Enshrined in the credo is the reason for the company's founding and a recognition of the people to whom it was responsible. The company literally chiseled it in stone, placing it in the lobby of its headquarters. It's been updated somewhat through the years as a living document. Today it reads:

> We believe our first responsibility is to the patients, doctors and nurses, to mothers and fathers and all others who use our products and services. In meeting their needs everything we do must be of high quality. We must constantly strive to provide value, reduce our costs and maintain reasonable prices. Customers' orders must be serviced promptly and accurately. Our business partners must have an opportunity to make a fair profit.

We are responsible to our employees who work with us throughout the world. We must provide an inclusive work environment where each person must be considered as an individual. We must respect their diversity and dignity and recognize their merit. They must have a sense of security, fulfillment and purpose in their jobs. Compensation must be fair and adequate and working conditions clean, orderly and safe. We must support the health and well-being of our employees and help them fulfill their family and other personal responsibilities. Employees must feel free to make suggestions and complaints. There must be equal opportunity for employment, development and advancement for those qualified. We must provide highly capable leaders and their actions must be just and ethical.

We are responsible to the communities in which we live and work and to the world community as well. We must help people be healthier by supporting better access and care in more places around the world. We must be good citizens—support good works and charities, better health and education, and bear our fair share of taxes. We must maintain in good order the property we are privileged to use, protecting the environment and natural resources.

Our final responsibility is to our stockholders. Business must make a sound profit. We must experiment with new ideas. Research must be carried on, innovative programs developed, investments made for the future and mistakes paid for. New equipment must be purchased, new facilities provided and new products launched. Reserves must be created to provide for adverse times. When we operate according to these principles, the stockholders should realize a fair return.

This is more inspiring, perhaps, than the early missions of Google or Facebook: "Don't be evil" or "Move fast and break things." The themes echoed in today's corporate mission statements are often too broad and too vague to provide much inspiration or guidance in a crisis. There are even online "mission statement" generators that churn out succinct platitudes and catchy, alliterative phrasing.

That was not enough for Robert Wood Johnson. He once wrote to his fellow business leaders that "industry only has the right to succeed where it performs a real economic service and is a true social asset." And he meant it, both in serving customers and in how he treated his workers.

By the time of the Tylenol crisis, Johnson had long since retired, but his legacy lived on in James Burke. Burke grew up in a small town in upstate New York, where he attended a strict Catholic high school. He later said that his school had contributed to his strong moral compass. After graduating, he joined J&J in 1953 as the manager of its Band-Aid product line. He eventually rose to lead the entire company.

Even without family leadership, the credo continued to exert an enormous influence over J&J's culture. It was Burke who launched the Credo Challenge, a series of discussions among company leadership about how it should be implemented from day to day.

But talking about values in the abstract is much easier than living them when the heat is on. What could Burke do in the Tylenol crisis? With no defined crisis management plan nor even a public relations department outside of the marketing division, Burke and his team were on their own.

In times of corporate crisis, business leaders tend to rely on narrow economic thinking—tallying up costs and benefits—to justify their decisions. Especially at large public companies such as J&J, most decisions are based on careful, quantitative analysis, with numbers leading the way. At times this sort of cold calculus comes across as heartless.

At times, in fact, it is. In perhaps the most widely cited case study of corporate indifference to human life, Ford Motor Company calculated that it would be more expensive to redesign its Pinto model to prevent it from exploding after minor accidents than it would be to pay off the victims' families after the fact: whereas it would cost the company $137 million to implement the design shift, it would cost only $50 million to pay for the estimated 180 deaths that would result by doing nothing. Ford ended up recalling 1.4 million Pintos in 1978, but it was too late to save the company from becoming a standard part of business ethics syllabi ever since.

Pulling Tylenol from every store in the country would have cost millions of dollars and would mean throwing away countless clean, safe bottles. Doing so would certainly have drawn more attention to the poisonings and further jeopardize the brand and maybe the company. That could have been the end of Burke's flagship product.

But Burke did not rely on financial modeling to tell him what to do. Instead, he relied on the credo, immediately pulling from the shelves every single Tylenol bottle in the United States—30 million bottles in all. He made the decision without even knowing the full cost, which would eventually reach $100 million. He later said that he hadn't even had to think twice. Why not? Because the credo was so clear and he believed so strongly in it. *We believe our first responsibility is to the patients.*

J&J cooperated with the Chicago Police Department, the FDA, and the FBI in all matters—except for the recommendation by officials to forgo a major recall. Concerned that it would prove that the poisoning could "bring a major corporation to its knees," the FBI and the FDA thought a recall would be an overreaction. But Burke's first responsibility was to patients, not to the FBI—so J&J launched the recall anyway, against its advice. The company also took the highly unorthodox step of inviting the press into its war room to observe the deliberations during that challenging time, including the much-feared Mike Wallace of *60 Minutes*.

Sure enough, Tylenol's leading market share dropped precipitously from more than one-third to below 10 percent. But within a year, J&J's openness with the media and its newly designed tamper-resistant packaging allowed it to restore trust. It soon reclaimed nearly all of its lost market position. Burke's response to the crisis earned him a place in *Fortune*'s Business Hall of Fame and won him the Presidential Medal of Freedom in 2000. In President Clinton's remarks about Burke, he said, "Thanks to him, our families are healthier, our communities are safer, our nation is stronger."

The J&J credo answered a question that few companies even bother to ask: Why does this corporation exist?

The first task of any leader should be to determine an organization's deeper purpose, beyond consideration of profit. It was this first task that Josh Silverman set for himself when he took the helm at Etsy in 2017.

ETSY AND THE CORPORATE CHARTER

"Why does Etsy exist?" Remembering back to when he had first become CEO, Josh Silverman thought a lot about that question.

The former chief executive of Skype, he had served in executive positions at American Express and eBay. He joined Etsy's board in 2016 and was named CEO in 2017. Brought in as an outsider, Silverman displays the zealotry of the recently converted. As measured by the Motley Fool's "ESG Framework," "his enthusiasm for Etsy is palpable in conference calls, investor days, and every interaction he has with shareholders." He saw leading Etsy as a defining role in his career. "I was patient and picky. This was my keystone."

Today, Silverman is the type of person who says "craftsmanship" and then corrects himself to "craftspersonship." In describing Etsy's headquarters, he touts its history as a former factory of the Jehovah's Witnesses. As one *Fortune* reporter described him, "He's a bald, bearded man who speaks in a soothing voice that any psychotherapist would envy. That trait came in handy amid the tumult surrounding his arrival."

The day before Silverman first addressed his employees, the company had fired eighty people, 8 percent of its workforce. It was the first big layoff in the company's history. Not much later, Silverman fired another 140.

As a place to connect craft makers with buyers online, Etsy had always been built around a sense of mission. And it had become a big business. As one reporter summarized in 2016, Etsy "sold $2.39 billion worth of block-printed whale posters, felted party banners, vintage Thonet dining chairs, and Liberty-print diaper covers hand-stitched by a grandmother just outside Bath."

But after a meteoric rise from its founding in 2005 to when it became a public company in 2015, the company had stalled. Growth had slowed, and it had remained unprofitable. Amazon launched Amazon Handmade in 2015 to compete. It was just another small business at the internet giant, but it was an existential threat for Etsy.

When Silverman's predecessor, Chad Dickerson, had taken the company public, its share price had closed the first day at $30. At their lowest point, nine months later, Etsy shares had lost 77 percent of their value. Shareholders were angry. "Nobody objects to having a good social mission unless and until it's at their expense," Silverman later said.

The shareholders' uproar and the board's decision to replace Dickerson as CEO became a sensation in the financial press. Dickerson had helped build the mission-driven business for nearly a decade. He had gotten Etsy certified as a B Corp, which is extremely rare for a company of its size. But then he had been replaced by an outsider with a more traditional business background. It seemed like a verdict on the entire responsible capitalism movement. According to *Bloomberg Businessweek*, "The Barbarians Are at Etsy's Hand-Hewn, Responsibly Sourced Gates." Here was this mission-driven B Corp that had gone public, now forced to give up everything it cared about in the name of profit. It became a cautionary byword among social entrepreneurs: "Remember Etsy."

"I'm not crying," wrote one Etsy engineer at the time, "I'm just allergic to capitalism."

Silverman inherited a manifesto that had been written by Etsy's long since departed founders when the company was formed, but he didn't believe it captured the right mission. *Why does Etsy exist?* It was a question he kept coming back to. His answer would determine whether Etsy would become just another shareholder-obsessed corporation or remain driven by something greater.

In some ways, it is a question that harks back to the very beginning of corporations. If you go back far enough, all corporations required a charter from the crown or state that stated explicitly what purpose they served and why they deserved the privileges of the corporate form.

In 1819, Supreme Court chief justice John Marshall described the way corporations owe their very existence to the state: "[The] corporation is an artificial being, invisible, intangible, and existing only in contemplation of law. Being the mere creature of law, it possesses only those properties which the charter of its creation confers upon it." From 1790 to 1860, state legislatures chartered more than 22,000 corporations.

Historically, government asked for something in return for the privileges of limited liability, legal personhood, and freely transferable shares. The corporate charter had to explain what purpose the corporation would serve and in turn how that purpose would serve society.

When the Granite Railway Company applied for a charter from the state of Massachusetts in 1826, it argued that "it would be of great public utility to establish a Railway" in order to transport stone from the inland quarries to coastal construction sites. But to achieve that social purpose, the railroad would require the "employment of a large sum of money, and that such a sum can only be obtained by extending the subscription among many persons." So the company asked the state legislature to grant it the privilege of incorporation.

That sort of justification was required. As Judge Spencer Roane of Virginia wrote in 1809, "Associated individuals [want] to have the privileges of a corporation bestowed upon them; but if their object is merely *private* or selfish; if it is detrimental to, or not promotive of, the public good, they have no adequate claim upon the legislature for the privilege."

That said, it was not solely for public interest that corporations were formed. At the 1896 meeting of the American Economic Association, Henry Carter Adams noted in his president's address, "A corporation . . . may be defined in the light of history as a body created by law for the purpose of attaining public ends through an appeal to private interests." It was as President Abraham Lincoln had described the US patent system: it "added the fuel of *interest* to the *fire* of genius." From the very beginning, corporations were meant to balance public and private purposes.

If there's no way for a corporation to have a profitable business model while serving a social purpose, maybe it shouldn't be a standalone, for-profit corporation. If there's no social purpose that fits a given profitable business model, maybe it shouldn't exist. Early in corporate history, this kind of thinking would not have seemed radical.

Charters were gradually replaced by "general incorporation codes" in the middle of the nineteenth century. Whereas charters had been granted by state legislatures, a general incorporation code was available to any company that filled out the paperwork and paid the fee. To be sure, general incorporation codes put an end to the political favoritism and tedium of issuing and amending charters that had plagued legislatures. But the shift had another effect: over time, people forgot that the corporation—with its unique privileges of limited liability and the rest—had always been meant to serve a deeper purpose.

All corporations still have articles of incorporation. Remarkably, they retain a vestigial structure that evinces the origin of the corporate form: in the articles of incorporation, Article IV states the corporation's purpose. But today, that purpose is almost always some form of the same narrow promise: "To do what is lawful in the state of Delaware." Corporations organize private interests efficiently and productively, but they are powerful cars without steering wheels. Charters were meant to be the steering wheels.

Registering a corporation with a real purpose can be done. We know because we've done it. Defining corporate purpose voluntarily at all corporations is a critical first step if capitalists are to reform capitalism.

Back at Etsy, that was the task Silverman was working on: defining the deeper social purpose that Etsy was meant to serve so that he could rebuild the company around that mission.

PROPOSAL 1: RECHARTER CORPORATIONS AROUND A DEEPER PURPOSE

Looking at Etsy's business model and stakeholders, Silverman determined that its core mission—the thing that it had been formed to do—was to create economic empowerment for its 2 million sellers. Eighty-seven percent of the sellers were women, 97 percent of whom were running their business out of their homes.

Its founding vision as "an anarchist artist collective" had never been a clear guiding light for making business decisions. Silverman's vision was of an army of independent artisans who would be able to make a living by expressing their creativity through handmade goods sold directly to consumers through Etsy's marketplace. That all fed into a new mission painted on the company's building and sent as stickers to its sellers: "Keep commerce human."

Just as J&J's credo specified patients to be the company's primary constituent, Silverman would refocus Etsy around service to its sellers by finding them more buyers. Nearly every all-hands meeting begins with a video of a seller telling her story. "We always start with the people we're here to serve," Silverman said.

To determine this purpose, he asked, "Where are we uniquely positioned to have maximum impact?" His answer was not something tangential, some flashy CSR project or corporate philanthropy. "The social mission has to be one and the same as its economic business," he said. "Our day job is our social mission."

Based on this new focus, he ended many initiatives that might have had some merit but did not contribute meaningfully to the core mission of economic empowerment for sellers. "What are the fewest things we need to do to succeed?" he asked. He eliminated the stand-alone corporate social responsibility team—known as the "Values-Aligned Business" team—and embedded its responsibilities into other people's jobs.

To measure progress in delivering on economic empowerment, Silverman focused on gross merchant sales, the total dollar value of the products that sellers sell through Etsy. With this in mind, he targeted initiatives such as sitewide sales and better search engine results. These are not radical ideas in themselves, but a clear purpose allowed Silverman to focus on them rather than get spread too thin. Before he had come in, there had been eight hundred business development initiatives at a company with barely that many employees. He cut half the initiatives.

By clarifying Etsy's purpose and relentlessly narrowing focus to the few things that would further it, he was able to reaccelerate growth on the platform. Gross merchandise sales surged in 2018, growing 21 percent over the previous year. Today, Etsy sees this as only the beginning, believing it has captured just 2 percent of its potential market. If that's right, it has a lot of economic empowerment left.

This is our first proposal for all corporations: Recharter around a deeper social purpose. Purpose is the most distilled form of strategy. It clarifies how a corporation should spend its time and resources. It aligns all actions around a common goal. And it motivates all stakeholders through a mission that is more inspiring than profit maximization. Health care companies should heal us. Education companies should teach us. Food companies should nourish us. Finance companies should enrich us. News companies should inform us as an engaged citizenry.

This, we believe, is the primary antidote to fiduciary absolutism. Purpose focuses managers on creating long-term value and reducing long-term risks. As we've now learned, increasing long-term growth and reducing long-term risk—difficult to define though risk is—should lead to a higher valuation multiple. In our experience, return and impact go hand in hand. They are mutually reinforcing. Unlike after-the-fact corporate social responsibility or socially responsible investing that is superficial, unambitious, and unfocused, the chartered enterprise harnesses the engine of capitalism to fundamentally align it with our broader social goals.

At the SOCAP conference in 2017, Jed Emerson warned the capitalist reformers in the audience against taking the "lowest path to the nearest point." It was a warning against the feel-good, quick-win solutions that play well in CSR reports and Super Bowl commercials but fail to address the rot at the heart of our corporations.

Over the last forty years, we've seen what happens to business when it follows the siren song of short-term profit. We can only imagine what rechartered corporations guided by the North Star of a deeper social purpose might accomplish. Rebuilding corporations along these lines will be no easy task, especially as so many investors and corporate leaders are still under the sway of fiduciary absolutism and wary of anything that smells of do-goodery. But this type of fundamental change is worthy of our highest efforts.

PROPOSAL 2: HOLD THEM ACCOUNTABLE TO THAT PURPOSE

For purpose to work as a guiding light for businesses, it must be deeply rooted in the governance of a company. It must be more than a vague mission statement to be bandied about at corporate retreats and printed in corporate social responsibility reports. Done halfway, it inspires cynicism without adding clarity. To be true to purpose, a company must set its targets against a set of key metrics that go beyond its financial statements. In addition, it must have a board that is both charged with and capable of ensuring faithful adherence to purpose.

"If you don't hold yourself accountable, you won't get an outcome," Silverman said. "People see whether you live it or not in the hard decisions." This is what we saw in the J&J crisis and what forms our second proposal.

Silverman believed that sustainability is a core part of delivering economic empowerment. It was not enough to say that he wanted the company to be sustainable; he needed to create a structure of accountability that would force Etsy to make difficult changes.

So he set a public target, putting a stake into the ground: he announced that Etsy would be carbon neutral by 2020. The building that houses Etsy's headquarters in Brooklyn, constructed before Silverman's time, was already a beacon of sustainability. But a vast majority of Etsy's impact comes from outside its doors. The company calculated that 98 percent of its carbon footprint came from transporting boxes from sellers to buyers.

Etsy held itself accountable for all the impact from its deliveries, even if it was not the one printing out the shipping labels or driving the boxes across the country. And so, in the winter of 2018, it became the first global e-commerce company to offset 100 percent of its carbon emissions from shipping. Once Etsy started notifying users in their shopping carts that the carbon would be offset, more buyers completed their purchases.

Etsy was able to achieve its goal of running a zero-waste operation globally by 2018, two years ahead of schedule.

Besides setting public targets, Etsy drives accountability through its monthly meetings and its annual shareholder report. In its monthly meeting—a multihour marathon session of seventy managers and hundreds of slides—every manager is expected to report not just on core operational and financial results but also on social and environmental metrics. "The point of the meeting," explained Silverman, "more than anything else, is accountability. And not just to me, but to each other." The company applies the same level of scrutiny to nonfinancial metrics as it does to financial metrics.

Rather than issuing separate financial and corporate social responsibility reports, Etsy issues an integrated report, so that all of the information is shared with all of the investors. It has even led the way on incorporating a thorough and auditor-reviewed SASB report in its 10-K SEC filing—a critical step we hope others will follow. It sets targets for its impact just as it does for financial growth. Its 2018 annual report set the goals of doubling Etsy sellers' economic output by 2023, doubling the percentage of black and Latino employees at the company by the same year, and powering its operations entirely by renewable energy by 2020.

In these ways, Etsy management holds itself accountable to the purpose it has designated for itself, just as J&J management holds itself accountable to the credo. Though there is still a long way to go for most public corporations to reach this level of accountability, we would push for them to go even further by explicitly empowering their board of directors to become the guardian of their purpose.

A board is charged with carrying out long-term, strategic oversight of a company and making sure that executives are delivering against their goals. It is meant to be the conduit between shareholders and the executive team, and it has the power to hire, fire, and compensate CEOs. The best corporate boards are highly informed, highly engaged in strategy, and clear about the purpose of their company. Sadly, many boards are stacked with retired grandees and faces friendly to the same senior executives they are meant to oversee. Many board positions today are ceremonial—the haunts of celebrity dilettantes or, worse still, friends of the CEO, ready with their rubber stamps.

To take one example: Dina Merrill served on the board of Lehman before its collapse. Merrill was a socialite and actress whose credits included the 1988 comedy *Caddyshack II*. She had certainly been *around* money during her eighty-plus years; her father had founded a large brokerage house, and the first of her three marriages had been to a Colgate-Palmolive heir. She served on the board of her father's company for eighteen years.

Merrill was one of ten board members overseeing Lehman Brothers. She retired months before the crisis, at the age of eighty-three. Among the board members left to navigate the bankruptcy, nine out of ten were retired, and four were seventy-five years old or older—remarkable for such a complex financial institution.

Board members' salaries are often north of $300,000 and sometimes over $1 million. You would think that that would get you a lot of focus and effort. But board members spend, on average, 245 hours per year on the companies they're meant to oversee—that's less than five hours per week. Some board members provide as few as eight days of service per year. They are often elected pro forma; many are nonexperts and can't be easily removed. Vacancies often go unfilled.

Board members are already liable for things such as negligence, but we think that their responsibilities should go further and include independent stewardship of the purpose embedded in a corporation's charter. The explicit mandate of corporate boards should be to balance long-term stakeholder interests in pursuit of that purpose. Doing so will serve shareholders in the long term.

The chain of communication that once existed among shareholders, employees, customers, and managers has become a garbled game of telephone. The only message consistently relayed is that of short-term profit. This enables what we've described as the Milgram-Nixon Syndrome, with companies doing things that shareholders would never do themselves. This happens because of the distance and intermediation between the two. Clear purpose and accountable stewardship must bridge that distance.

However, even if this is done perfectly—with a clear charter and an enlightened board—it will all be for naught if the corporation is constantly forced to choose between profit and purpose. Indeed, as we'll see, even J&J has drifted off course. Unless a company's business model and culture align with its purpose and the prosperity of all stakeholders—shareholders, managers, employees, and the rest—trade-offs will abound. Sooner or later, the corporation will fall prey to its worst incentives.

PROPOSAL 3: ALIGN THEIR PURPOSE WITH THEIR PROSPERITY

When James Burke was awarded the Presidential Medal of Freedom in 2000 for his outstanding leadership, Alex Gorsky was a vice president of J&J's subsidiary Janssen Pharmaceutica. That year, Gorsky's team managed to exceed its $1 billion sales target for the antipsychotic drug Risperdal.

Sales to adults had plateaued, so Janssen grew by marketing Risperdal "off label" to children and the elderly. However, a study conducted by J&J the same year found that Risperdal led to a significant increase in prolactin levels in young boys. Prolactin is the hormone responsible for breast development, and "it can cause boys to grow large, pendulous breasts; one boy developed a 46DD bust." J&J ignored the study for nearly ten years.

Eventually, it was forced to pay more than $2 billion in penalties and settlements. (To put that number into context, Risperdal has accumulated more than $30 billion in sales). It's hard to imagine

Gorsky walking past the credo in the lobby every day while overseeing that division. But in the post-Burke era, the credo has become a monument instead of a living charter. Despite all this, Gorsky is now CEO and chairman of J&J, earning a total of $50 million in compensation in the last two years alone.

When one reporter asked eight of J&J's board members about their promoting Gorsky, given his role in Risperdal, each of the independent board members responded with one voice: speak to J&J's corporate public relations department.

More recently, in 2018, J&J was ordered to pay more than $4.5 billion in compensation and damages over claims that its talcum powder products caused ovarian cancer. Evidence suggested that some executives had been aware of the potential link between talcum powder and ovarian cancer for decades.

Below purpose and effective governance sits the business model. Unless the business model is rebuilt to serve the company's purpose, stories such as J&J's fall from grace will be all too common.

Other companies have done better to align their purpose with how they make money. Kaiser Permanente is a unique health care company in that it is an insurer and provider in one. By being a patient-member of Kaiser, you are essentially paying it to keep you healthy. If Kaiser can provide sufficient preventive care to keep you out of the emergency room, the company makes more money. It does not profit from pushing patients through expensive, ineffective treatments. Nor does it profit from withholding care that has been proven effective. As a result, Kaiser has become both highly admired and highly profitable.

Or consider some of the most successful coding academies. Rather than charge up-front tuition, companies such as Vemo Education and Lambda School take a portion of their graduates' salary for a limited time. If their students can't find work after graduating with income above a certain threshold, the schools make no money. If they find high-paid coding work the day after they leave, the schools make lots of money. Their purpose is reflected in their business models.

John Mackey of Whole Foods described his perspective as follows: "If you look for trade-offs, you will always find them." The flip side is that if you look for areas that are win-win, you will find those, too. In this debate, you'll always find what you're looking for. You have a choice: set up business models that are full of trade-offs, or set up businesses that are full of win-wins. There are a million ways to build a company; you get to decide.

Which brings us back to Etsy. For all that Etsy has done over the past five years under Silverman, the company has not gone uncriticized. "Etsy had the potential to be one of the truly great ones," said one employee early in Silverman's tenure. "But it looks like they are cutting anything that's not essential to the business. This is a cautionary tale of capitalism."

Sellers make a huge commitment to Etsy by building their businesses on its platform. So it's not surprising that they've always been wary of Etsy's motivations, especially when they feel that the company's success is coming at their own expense. "The distrust started an hour after the website went live," Silverman said, "and has never stopped."

The problem is that Etsy's business model can be at odds with sellers' best interests. For example, Etsy recently raised its fees from 3.5 percent to 5 percent on each item sold. Silverman argues that higher fees enable the company to invest in building a better platform and that the sales speak for themselves. But others see misaligned incentives. Silverman can sound ruthless when he talks about this. "It's not our job to be popular," he said. "It's to be effective." He continued, "That doesn't mean pandering to everything everyone wants. It means being data driven." For Silverman, being a leader committed to a deeper purpose sometimes requires making unpopular decisions. In the end, it's all about balance.

So far, this seems to be working for Silverman and Etsy. Sales are up, he's increased the employee head count, and Etsy's on track to hit its environmental targets. The Motley Fool rated Etsy 9 out of 10 on its ESG Framework, noting that "Etsy seems to be balancing profitability, scale, and strong ESG principles." Etsy has held off Amazon Homemade from stealing its market.

As a result, shareholders have become more optimistic about Etsy's long-term prospects. After falling by 62 percent before Silverman was announced as CEO, the stock price under his leadership has nearly quintupled. For those locked into a trade-off mentality, this is worrying; they feel that it must have been at the expense of other stakeholders. For those who see a purpose-driven business holding itself accountable to its goals and aligning its purpose with its prosperity, this is an encouraging success story.

According to Silverman, shareholders thought "the results were terrible *because* of the mission, and that's a terrible way of looking at things." But many observers remain skeptical: "There's only so much wiggle room as a public company," one early employee told the *New York Times*. "If you really want to build a company that works for people and the planet, capitalism isn't the solution." In the words of *The Cut*: "Etsy Wants to Crochet Its Cake, and Eat It Too." Silverman believes Etsy has to hold itself to an even higher bar financially because of its mission. And he's trying to answer the skeptics with results.

For Etsy, these are still early days. The company could end up like J&J, where talk of its social responsibility now comes only in the context of recent scandals. Or it could end up like Unilever, still winning accolades more than a century after its birth. It will be up to Silverman and his successors, the people on his board, the employees of his company, and his shareholders to decide. Just as a sports team can never rest on its accolades but must start each season anew, so, too, will Etsy have to prove its value again and again—to its sellers, its buyers, its shareholders, its employees, and society.

REFLECTING ON ETSY

"For better or for worse," said Chad Dickerson two years after he was ousted as CEO at Etsy, "and probably for worse—if you consider me getting fired—I often made decisions that were not in the shareholders' interest."

Dickerson had joined Etsy as its chief technology officer in 2008, when the company had more aspiration than discipline. As the company was about to make its third CEO change in as many years, Dickerson put himself into the running. In 2011, the board named him CEO.

"There was a lot of talk about values, but the company didn't even have a stated list," he said. "There was no way to measure how we were doing even if we had stated values." Dickerson assembled a team across tenures to create a new values statement. He began tracking ESG metrics just as he did financial performance. He got Etsy certified as a B Corp, eventually pushing the company from just barely crossing the certification threshold to reaching the top 10 percent of B Corps globally.

"To be honest, one of the things that annoys me about the Etsy narrative is that it was really undisciplined and then they brought in a CEO"—his successor, Josh Silverman—"who was really disciplined." In Dickerson's eyes, Silverman's work represents less a turnaround than a continuation of the company's previous successes.

But Dickerson worries that some of Etsy's recent moves overwhelmingly serve shareholders. "If you look at the conflicts between stakeholders, the evidence is already there. The shareholders are going to be primary," he said. He pointed to the connection between Etsy's raising its fees and the share price going up: "It's not a zero-sum game, but everybody does not win equally."

Dickerson doesn't deny the rough patch Etsy hit after it went public, and he gives Silverman credit for turning its performance around. But he was surprised that the narrative around his termination was based on whether Etsy was too focused on doing good. It reveals a double standard in the corporate world: when a purpose-driven company does poorly, it's because of its purpose; but when a profit-focused company does poorly, well, it's something else. When times get tough, investors' conviction that doing good is good for profits suddenly disappears. The data back Dickerson up: CEOs who focus on corporate social responsibility are 84 percent more likely to be fired when financial performance is poor.

Though he's proud of the work he did at Etsy, Dickerson had wanted to do more to lock in its mission forever. "When I look back at my time at Etsy, that's one of my biggest regrets: that we didn't get to the point where we could write the mission into our charter."

Without a charter that reflects its social purpose, can Etsy stay true to its mission? "They could do it or they could not do it," said Andrew Kassoy, the cofounder of B Lab, "but either way, they're not accountable." Kassoy argued that to create real accountability, we need to change the relationships between corporations and their investors. He pointed to the example of Whole Foods, which was acquired by Amazon in 2017. John Mackey, who had led Whole Foods with a true sense of mission for decades, had been forced into the sale by the corporation's shareholders. When he wrote his book *Conscious Capitalism* in 2013, he didn't believe it was necessary to change a corporation's legal form or governance to be stakeholder oriented and mission-driven. But when he found himself in a fight with his investors? "Boy oh boy oh boy, did I wish we were a B Corp," he said in an interview after the sale. He wanted a legal basis to defend the company's mission.

Reflecting on both Etsy and Whole Foods, Kassoy said, "It's just too much to ask benevolent CEOs—even if they're well meaning and believe in it—to act differently when all of the other incentives in the system are the same."

When Dickerson was still CEO of Etsy and its stock was at its lowest point, he made an unpopular decision—unpopular, at least, with shareholders. Etsy was still a young company with a young workforce, and many of his employees were beginning to have families. Dickerson wanted to institute a more generous parental leave policy. He remembered sitting in the boardroom as people argued whether it was the right time for such a move, given where the stock was trading. But Dickerson knew his employees were going to have kids whether the stock was up or down. So he pushed it through: six months fully paid leave, regardless of gender.

The cost of a program like that is tangible. It shows up on the income statement each quarter as an expense. The long-term benefits to Etsy are intangible, though they help explain why 56 percent of Etsy's employees are women, 67 percent of its executive team is women, and 38 percent of its tech roles are filled by women—when the industry average is 20 percent. You can't calculate these benefits precisely, just as Silverman can't calculate the benefits of all the social and environmental projects he has committed to and built on since taking control. Both leaders are believers in the power of business to serve a deeper purpose, and both have done their best to find the right balance in doing so.

But did Dickerson really have to change the parental leave policy when the stock was at its all-time low? He laughed. "It probably was not the smartest thing to do in terms of shareholders. Honestly, I didn't care. I still don't care." Dickerson lives in Brooklyn, where he sees many of his former employees and their families. They benefited from the policy, and many remain committed workers at Etsy. He sees how happy they are. "I like to think we had something to do with that."

REPURPOSING CAPITALISM

If corporations are big enough and influential enough to cause our problems, they should be big enough and influential enough to solve them. Boeing buys billions of dollars' worth of product from tens of thousands of suppliers. Walmart employs 2.3 million associates—more than the population of fifteen US states. Facebook has 2.2 billion users worldwide—nearly 30 percent of the world's population. The Fortune 500 is collectively worth *$23 trillion,* far outweighing the $428 billion Americans donate to charity each year and even the $4.4 trillion budget of the federal government. The decisions these corporations make in Washington, Arkansas, and California boardrooms have tremendous power. When giants walk, the world shakes.

Imagine all our companies responding to crises as J&J did with the Tylenol poisonings. Or driving sustainability like Unilever, nourishing us like One Mighty Mill, empowering us like Earnin, informing us like *Half Moon Bay Review,* or "keeping commerce human" like Etsy. At their best, these companies approximate our solution: corporations driven by a deeper purpose, holding themselves accountable to that purpose, and building their businesses around that purpose.

This is the $23 trillion solution. It unleashes the power of our biggest corporations to nourish us and heal us, enrich us and improve us. Repurposing our corporations unleashes the power of the private sector to attack our biggest social and environmental challenges.

Corporations were once regularly evaluated against their social purpose by the state legislatures that chartered them. If they violated their charters, they were said to have acted *ultra vires,* or beyond their powers. State legislatures could revoke their charter. States no longer exercise that power, but we as a society still do. This is what the Divest Harvard movement seeks: to take away oil and gas companies' licenses to operate; effectively, to revoke their charters.

In a smaller way, it's what's happening at each of the companies we've profiled that have fallen out of step with the values of our society: Kraft Heinz, Kellogg's, Digital First Media, traditional payday lenders, and now J&J, among others. However, the biggest risk to our society today is not that we revoke the implicit charters for individual companies but that the bad behavior of companies in general pushes us to revoke the charter for all of capitalism—to say that capitalism itself has acted *ultra vires.*

As we've outlined, we think there is a better way that preserves capitalism's creative engine while steering it to loftier planes. There are critical roles for government, nonprofits, labor unions, universities, and civic institutions to play. They've often been necessary counterweights to corporations and necessary actors in pushing corporations to do more. But without rebuilding corporations around a deeper purpose and eschewing the broken ideology of fiduciary absolutism, all other efforts will be insufficient.

The basic plumbing is in place, but it's been corroded by decades of the profit maximization dogma. We don't need to tear down the house, we just need to repair the pipes. Unleashed by the tremendous engine of capitalism, corporations have delivered unfathomable growth. Unbounded by any sense of purpose beyond profit, they've delivered social and environmental destruction. Our corporations can deliver on a deeper social purpose. And as we'll see in the next chapter, we can help them do it.

9

CITIZEN CAPITALISM

REBUILDING TRUST AND REALIZING THE COMMON GOOD

Australia was set to be the next victory in its quest for world domination. It would fall just like every other country, capitulating to the empire slowly creeping to nearly every corner of the globe. With hundreds of millions of dollars at its disposal, a master plan that included detailed maps, and a team of hundreds, it was only a matter of time. Australians would succumb just like everyone else: they would become Starbucks customers.

Starbucks entered Australia in 2000 with audacious plans. After growing from 84 stores in 1990 to more than 3,500 stores globally just ten years later, Starbucks viewed international expansion as its path to new profit. By that time, there were so many stores in the United States that the *Onion* joked that the chain was "opening its newest location in the men's room of an existing Starbucks."

Between 2000 and 2008, Starbucks opened eighty-four locations in Australia, trying to make its coffee shops as ubiquitous in Melbourne and Sydney as they were in New York City and Los Angeles. In the company's annual report to shareholders, the leadership was sanguine about the future, anticipating that it would soon dominate the market, hawking Frappuccinos and Caramel Macchiatos to the Antipodes.

For the existing Greek and Italian immigrants who had built Australia's coffee culture since the 1950s, it was a terrifying development. They had brought coffee to Australia, building hundreds of independent cafés serving the beverage as they did in their home countries: espresso, cappuccino, and the peculiarly named flat white. Cafés in Australia had evolved into important informal meeting places, often starting as open-air breakfast joints in the morning and gradually transforming into casual watering holes by evening. Their independent and informal feel provided a respite from the hectic world outside.

And now those independent cafés would be forced to compete with a global behemoth. The future did not look good.

This is an all-too-common story in the history of global capitalism. Barnes & Noble destroys the local bookstore. Walmart destroys the small-town retailer. Amazon destroys them all. The cafés of Australia would put up a fight, as the little guys always do. They would try to make their loyal customers stay loyal, matching prices and selection where they could. But eventually, the inexorable forces of the free market would replace them with identical replicas of coffee shops in the United States—just as in China, Japan, the United Kingdom, and the seventy or so other countries to which Starbucks had taken its winning formula.

In the grand scheme of US businesses, Starbucks is one of the best. It focuses on sustainable products. It treats its workers well, providing higher wages, education credits, and a path to advancement. When former CEO Howard Schultz speaks about Starbucks, his words are inspiring. "I set out to build a company that my father, a blue-collar worker and World War II veteran, never had a chance to work for," he said. "Together we've done that, and so much more, by balancing profitability and social conscience, compassion and rigor, and love and responsibility."

But despite the company's immense resources and friendly practices, something remarkable happened in Australia: Starbucks failed.

Over the course of the ensuing decade, Starbucks accumulated $105 million of losses and was forced to shut down 70 percent of the stores it had opened. Those that remained catered mostly to tourists.

"Starbucks underestimated the rich coffee culture in Australia," said Nishad Alani, one of the Starbucks executives charged with scaling down the Australian operation. "Aussies enjoy their cafés just the way they are." They already had a thriving café culture with high quality java and a few national brands: Gloria Jean's, the Coffee Club, and Wild Bean Cafe. Market forces notwithstanding, the Australians didn't want Starbucks replacing their café culture, and Starbucks was forced to retreat.

Australian café owners enjoyed a final coup de grâce nearly twenty years later when Starbucks announced its newest global beverage: the Australian flat white. Rather than selling Americanized coffee to Australians, Starbucks was now selling Aussie coffee to Americans. One Australian critic offered his judgment of the drink: "For Australians who want to come to America and pretend they've never left home, Starbucks is offering a useful service." Turning the tables on Americans, he asked, "But will it catch on with the locals?"

The story reveals something important: capitalism is nothing more than the sum of our individual choices. Our economy is no more moved by an invisible hand than a Ouija board is moved by invisible spirits. It's just us.

We've seen how corporations, institutions, and governments can become divided against themselves. Many people suffer from the same affliction, leading economic lives that are at odds with their values. Markets are only as amoral as we are. The decisions we make as buyers, workers, savers, and voters can bend corporations to our will.

Throughout the book, we've seen a growing trend of people ending the false separation between their moral and economic lives: consumers bringing their values to what they buy, employees to where they work, and investors to where they put their money. In this chapter we'll see how each of these—as well as how we exercise our vote—is pushing capitalism toward reform. We'll see how it's brought vegan Whoppers to Burger King, an employee walkout at Wayfair, and a Cambrian explosion of new ESG-focused mutual funds. We'll see how we can build on these efforts to achieve even more.

Against the shibboleths of fiduciary absolutism, we offer a new vision: citizen capitalism. It's an economy made up of corporations chartered around a deeper purpose and individuals living their values in each of their roles: citizen buyer, citizen worker, citizen saver, and citizen voter. It's a society in which we each take responsibility for our economic power and ensure others take responsibility for theirs.

Ultimately, our affliction is cultural: we must reject the ideology that says private vice makes for public virtue, that corporations can aspire to nothing more than maximizing profit. Our business leaders must take the first step, but the rest of us should push them to do so. And we should meet them halfway when they do.

We've talked about government. We've talked about charters and corporate governance. It's time to talk about us.

CITIZEN BUYERS

In 2019, Burger King rolled out the Impossible Whopper, a new vegan patty that mimics the taste and texture of ground beef. It even bleeds like meat. Burger King initially released the Impossible

Whopper at fifty-nine restaurants around St. Louis. It's now selling it in more than seven thousand locations nationally. And it's not alone; McDonald's, White Castle, and Carl's Jr. are all offering vegan alternatives.

Meat consumption has a major impact on our environment. Every year in the United States, 10 billion land animals are killed for food. Together, these animals account for more than half our water use and a sixth of greenhouse gases. The creation of pastureland on which to raise cows is a major contributor to deforestation in places such as the Amazon.

Burger King's improbable decision to introduce the Impossible Whopper came from an unlikely place: 3G Capital, the Brazilian private equity firm that owned Kraft Heinz and attempted to take over Unilever.

Who at 3G is the vegan animal rights activist? It doesn't matter. If the customer cares, the company has to care. A competitive economy operates in infinitely iterative feedback loops. Companies must compete with one another to match our preferences. It's part of what makes dealing with a monopoly—the cable company or the DMV—so frustrating: you're stuck. When forced to guess, people think that corporations have profit margins averaging 36 percent. That's 4.5 times too high. The average profit margin is in the mid-single digits. That's thanks to competition.

Today, 55 percent of consumers say they would pay more for a product from a company that had a positive social impact. This is up from 45 percent just a few years ago. As a result, companies are increasingly competing to show consumers just how positive their impact is.

Unfortunately, it's not easy for consumers to express their values in all the thousands of economic decisions they make each year. As we have seen, two in five Americans can't come up with $400 for an emergency. They can't afford the luxury of eating only organic food or shopping at B Corps. And for those who can afford to, often they don't. The book *The Myth of the Ethical Consumer* revealed that we express more of our values in surveys than we do in the supermarket aisle. As one chief executive told researchers off the record, "Almost all of our customers are interested in what we can do to clean the environment and other stuff. You can tell it's one of their core values . . . until you get to price."

Yet 70 percent of our economy is made up of consumer spending. For those who can afford to vote with their wallets, where should they start? Fortunately, consumers don't have to be saints to include their values along with the other considerations of cost, quality, and convenience.

Of all the countless buying decisions we make each year, a handful make up the majority of our spending. The average household spends 33 percent of its money on housing, 16 percent on transportation, and 13 percent on food. That's almost two-thirds of all spending in just three categories. Let's look at each.

Within housing, the environmental impact of where to live far outweighs most subsequent decisions. LEED-certified buildings use 11 percent less water and 25 percent less energy than buildings that do not meet LEED standards, and they emit 34 percent less carbon. In commercial real estate, occupancy rates in LEED-certified buildings are now more than 10 percent above those of other buildings, and premises sell at a significant premium—because that's what customers are demanding. As consumers push for more sustainable living arrangements, the housing stock will slowly shift to meet that demand.

Transportation is similar. All cars advertise their fuel economy. Consumer demand has pushed electric vehicles into the mainstream. By 2040, more than half of new cars could be electric, and they may become less expensive than gas-powered vehicles as early as 2022. We can devote more attention to housing and transportation decisions because they are decisions we make so infrequently. Drivers keep a new car for six years on average. We decide how we'll commute once and then do it the same way every day for years.

As for food, grocery stores are a good example of where consumers can aggregate decision making to a trusted source. Rather than figure out which of the thousands of brands in a grocery store are ethical, we can pick a grocery store we trust to pick the right brands for us.

Thrive Market, for example, is an online grocer specializing in healthy, eco-friendly products. Its goal is "making healthy living easy and affordable for everyone." It does this by creating an approachable experience for those who care about wellness but have no idea what "phthalate free" even means. So Thrive tries to provide transparency without overwhelming its members. This means providing products that are—deep breath—Fair Trade Certified, Non-GMO Project Verified, BPA free, organic, and compostable without requiring members to research every certification extensively. For each category, Thrive does it for them. It tries to be on the cutting edge of what's healthiest and most sustainable, which often means going back to smaller, more traditional vendors.

Thrive has 600,000 members and is growing faster than ever. In addition to pioneers such as Thrive and Whole Foods, more mainstream retailers are beginning their own efforts to capitalize on health and sustainability trends. Walmart and Target have both been investing in organic and natural foods, and Walmart is now the largest organic foods retailer in North America.

Consumers are also amplifying the power of their decisions by using their voices. Corporations, on edge to stay in consumers' good graces, are becoming more sensitive to feedback, especially when it's tied to a social or environmental cause. Cell phones and social media have created an army of vigilantes all too ready to hold corporations to account. Slacktivism—activism from behind the comfort of a keyboard—gets a bad rap, but when used in conjunction with other tools, it plays a meaningful role.

And of course, for those focused on sustainability, the most environmentally responsible new dress is the old dress. Fast fashion has filled our wardrobes with disposable outfits, worn for a night and forgotten. It's the realization of a 1955 advertisement in *Life* magazine that promoted "throwaway living" as an answer to endless cleaning. Why wash a plate when you can use a paper one and then throw it away? Since that time, the average American home has tripled in square feet, even as the average family size has declined. Against this megatrend, the "slow clothing" movement has developed to reduce the footprint of fast fashion. Sustainable fashion is now pushing back, trying to get people to buy high-quality clothing, sustainably made, that lasts for years—not days or weeks.

Individually, these seem like small gestures. But it is thanks to a series of small gestures that Australians kept their cafés and Burger King customers got the impossible—at least as far as Whoppers go.

CITIZEN WORKERS

In June 2019, 547 employees of the online home goods company Wayfair sent a petition to the CEO. When they didn't get an adequate response, they staged a walkout and made national news. Months before, Wayfair had signed an agreement to sell $200,000 worth of bedroom furniture to the nonprofit BCFS Health and Human Services (formerly Baptist Child and Family Services), which was operating migrant facilities for the Department of Health and Human Services.

It may appear incongruous that a company selling a "heart wall decor" for $39.99 and a "wellness series desk lamp" for $113.99 was supplying beds to a detention center, but, as Wayfair management wrote to the protesting employees, "we believe it is our business to sell to any customer who is acting within the laws of the countries within which we operate."

That did not sit well with the employees. They demanded that

Wayfair have "no part in enabling, supporting, or profiting from" the "detention and mistreatment of hundreds of thousands of migrants seeking asylum in our country." They wanted a new code of ethics that "empowers Wayfair and its employees to act in accordance with our core values."

From the point of view of Wayfair cofounder Steve Conine, sales like these are all part of doing business. The *political* arena is "really the appropriate channel to try and attack an issue like this." But Wayfair itself? "To pull a business into it—we're not a political entity. We're not trying to take a political side." Ultimately, management's response to the walkout was to donate $100,000 to the Red Cross. But the walkouts and the publicity demonstrated something important: that workers have a voice.

Protests like these are growing in prominence, as employees are now holding their companies to a higher moral standard than "if it makes money, do it." Google was forced to cancel a contract with the Department of Defense in 2018 after thousands of employees protested. "We cannot outsource the moral responsibility of our technologies to third parties," the employees wrote in a letter to Sundar Pichai, the CEO of parent company Alphabet Inc. "Google's stated values make this clear: *Every one of our users is trusting us. Never jeopardize that.*" Similar employee protests have erupted at Microsoft over weapons technology, Amazon over climate change, and the big-data analytics firm Palantir over its work with the Immigration and Customs Enforcement agency.

Employee activism highlights the most important economic transaction most Americans make every year. It's not what they buy but what they sell: their labor. After all, the economy is nothing more than the sum of all of our career choices. Everything bought must be sold; everything sold must be built. As a general rule, whenever we spend a dollar, we must first have earned it somehow.

Sixty-seven percent of employees say they would prefer to work for a socially responsible company. Keyvalues.com, Idealist.org, and

80,000 Hours have become popular resources for job seekers to find mission-oriented work. Workers with the luxury to choose their job can use their voice to push their values. If potential hires refuse a job because they think the work environment is toxic for women, they should tell the company why. And they should tell it to the most senior employee they can.

Those who are already in jobs are feeling more empowered to push for the things they care about, calling out things they disagree with and advocating for what they believe in. Nearly three-quarters of US employees believe they can make a difference by speaking out against controversial issues at work, and nearly half of millennials claim they already have. In well-functioning organizations, these sorts of discussions can rise up to the top and are often addressed. After all, managers and boards are now seeing the risk when they let these needs go unaddressed. Just ask the executives at Wayfair.

CITIZEN SAVERS

In chapter 6, when we profiled the activist hedge funds now focusing on ESG, we left out one thing. At the end of the day, they are selling a product, just as much as Unilever sells soap or Gannett sells newspapers. They are investing money that's been entrusted to them by what are called "limited partners"—whom we might just call "customers." All investors—Jeff Ubben at ValueAct, Cliff Robbins at Blue Harbour, and the rest—are subject to what those customers want.

And, increasingly, those customers want to invest responsibly. Though less than half of baby boomers say ESG factors are important in investing, 65 percent of Gen Xers and 87 percent of millennials do. Because of regulations, individual investors cannot invest in hedge funds. But they can invest in mutual funds, and mutual funds are responding to changing preferences. By one computation, more than $12 trillion in the United States is invested in funds that aim to produce social and environmental benefits as well as financial returns. This has grown fourfold from a decade ago.

Several investment firms are now providing products specifically geared to socially minded investors. MSCI rates 32,000 mutual funds on an AAA to CCC scale, based on the environmental, social, and governance ratings of the stocks the funds hold. As You Sow provides online tools to find investment funds that match your values. They have funds focused on combatting climate change, achieving gender equality, or reducing gun violence. JUST Capital manages an investment fund that roughly tracks the Russell 1000 but includes only the top 50 percent of companies as ranked by their ESG performance.

Socially responsible investing is a nascent effort, especially for individual investors. Most investment funds still compete entirely on the basis of historical performance and low fees. But as individual investors continue to demand it, the sector will continue to develop, competing with more and better options for investors to align their values with their savings.

Beyond retirement accounts and stock portfolios, people have a choice of what banks or life insurance companies to use. Banks use the money in our savings and checking accounts to lend to others. Life insurance companies invest our premiums. These institutions are also increasingly being forced to compete to reflect our values. Triodos Bank, based in the Netherlands, has built its entire business model around pushing the banking sector to be more transparent, diverse, and sustainable, both through its own operations and in whom it chooses to finance.

And, of course, we are stakeholders of many institutions with financial resources invested on our behalf. Divest Harvard recognized that the Harvard endowment is invested partly on *their behalf* as students and demanded that it be invested according to their values. In a way, it's surprising that people still donate to their alma maters without asking how their money will be invested.

As younger generations earn and inherit the bulk of the world's wealth, their preferences will determine how their money is saved and invested. The financial world will ignore their changing preferences at its own peril.

CITIZEN VOTERS

Electric city buses in many major cities now bear the banner "No Emissions." Here's the only problem: it's not true. Or at least not *strictly* true. Yes, there are no emissions coming out of the back of the bus. But what's powering all those electric vehicles? If you drive a Tesla in West Virginia, you are really driving a coal-powered car. If you drive a Tesla in Oregon, it's water powered.

Electric or hybrid cars are more efficient users of power than those that run on gasoline, but it matters just as much—often more—how we *get* our energy as how we *use* our energy. The decision between an energy-efficient light bulb and a traditional light bulb is one one-trillionth as important as the decision on how to power our cities.

Ned Hall, the Harvard professor helping to lead the Divest Harvard campaign, said, "We've been trained too much to think about ourselves as consumers first and citizens second." So we mistakenly think that climate change is something we can address with lifestyle changes alone. That's not true. No amount of non-GMO corn chips, LED light bulbs, and reusable shopping bags will fix what's broken in our world. Changing the way we shop should not replace changing the way we vote.

Indeed, in a democracy, we are both the regulator and the regulated. Our republican form of government means that we make a few high-level decisions once or twice a year, entrusting our representatives to carry out our wishes.

So our first job as citizens is to cast a ballot. In the 2016 presidential election, only 56 percent of the voting-age population in the United States voted. That put the United States twenty-sixth out of thirty-two developed countries, somewhere between Slovakia and Slovenia. A higher proportion of Americans say that high turnout in presidential elections is "very important"—70 percent—than actually turn out to vote. Even if we vote with our wallets—and our pay stubs and our deposit slips—we still have to vote with our ballots.

In this book, we've seen how poorly our financial system serves the underlying interests of most shareholders, who are diversified, long-term savers who care about far more than just stock market return. The same is true of our political system, where broad agreement among the population can still lead to inaction in Washington.

Rather than focusing on hot-button issues such as health care or gun control, some organizations are focused first on fixing the gears of democracy. Leadership Now, founded by Daniella Ballou-Aares, is organizing a bipartisan effort to end gerrymandering and institute ranked-choice voting. It believes that changes like this will help our representative government to represent us better. Others are focused on getting corporate money out of elections or constraining the nearly $3.4 billion lobbyists spend each year.

Is it redundant to say *citizen* voter? Well, there are two ways to think about voting in a democracy: vote for your self-interest or vote for the common good, something the Swiss philosopher Jean-Jacques Rousseau described as "the general will."

Empirically, it appears that most Americans do not just vote for their economic self-interest. Though the rich skew Republican, 40 percent of those making over $200,000 a year voted for a Democrat in the 2012 election. Though the beneficiaries of welfare programs skew Democrat, 30 percent of those making under $25,000 a year voted for a Republican. Many Americans regularly vote against their own economic interests. Maybe it's because they don't understand what's in their self-interest. Or maybe it's because they care more about the country over the long term than about their tax rate or benefit check today.

Rousseau was aware that people can be selfish. But he argued that in a democracy, we should all have a loyalty to the common good that supersedes any personal interest. When we form a country, we create something bigger than its individual members. When we vote *as citizens*, we do so in the interest of that bigger project rather than our own narrow self-interest.

Being a citizen voter means voting for the common good, recognizing that we are all served best by serving one another. The same is true for citizen buyers, citizen workers, and citizen savers as well; in all cases, we are reflecting our broader public values in our private decisions.

Fiduciary absolutism is an adversarial mindset. It's about shareholders against stakeholders, government against corporations, private vice against public virtue. It divides us against ourselves as a country, as corporations, as institutions, and as individuals.

Citizen capitalism is a cooperative mindset. It's about working together on the common project of our economy for the common good of our society. It's about corporations that serve a deeper purpose and, ultimately, about citizens who live their values.

RESTORING TRUST WITH ACCOUNTABILITY

A scorpion asked a frog to carry it across the river. The frog hesitated. The scorpion insisted it would not sting the frog. "If I did," the scorpion argued, "we would both drown." Assured by this logic, the frog allowed the scorpion to get onto its back and began to swim. Midway across the river, the scorpion stung the frog and they both began to drown.

"But why?" asked the frog. The scorpion replied, "Because it is my nature."

In an economy so dominated by fiduciary absolutism, so broken by inequality and climate change, so diseased by selfishness and greed, something about the adversarial mindset feels right. It's a brutal world out there. Trust no one.

Trust no one. "A social organism of any sort whatever, large or small," wrote the philosopher William James, "is what it is because each member proceeds to his own duty with a trust that the other members will simultaneously do theirs." Without that trust, "not only is nothing achieved, but nothing is even attempted."

Three in four people do not trust business leaders to make the right choices on social issues. Four in five people do not trust business leaders to tell the truth and make ethical decisions. We don't even trust one another. Over the last generation, the proportion of American teenagers who believe that most people can be trusted has collapsed by 40 percent.

An underlying premise of all economics is that trade can leave everyone better off—that working together leaves everyone with more. That's why we trade. That's why we create companies. Distrust is corrosive. Through distrust, we see our economy as a zero-sum competition where someone can get ahead only by pushing someone else behind. The frog trusted the scorpion, and look what happened.

This is why Divest Harvard will settle for nothing less than divestment. It's why we resort to government intervention, even if it crowds out private responsibility. It's why corporate social responsibility and impact investing have been met with so much cynicism, even when they are done in good faith. If a corporation's highest purpose is increasing the next quarter's profit, we are justified in our wariness.

"Trust is really what keeps me up at night," said Halla Tómasdóttir, who leads a coalition of business leaders at the forefront of the fight to save capitalism, "because that's essentially the basis of our social contract. We cannot rebuild our broken social contract without rebuilding trust."

Tómasdóttir's organization is called The B Team. Rather than following Plan A, according to which companies seek only profit, The B Team seeks to catalyze a "movement of business leaders driving a better way of doing business for the wellbeing of people and planet." That's Plan B.

"Today, amidst rising temperatures, inequality and broken trust we are seeing the world demand not just courageous leadership, but also greater accountability," The B Team wrote. "Leadership that only values shareholders or financial performance must be replaced by principled leadership that holds itself to a higher standard and to all stakeholders."

The B Team was cofounded by Virgin Group Chairman Richard Branson and Jochen Zeitz of Puma, and it includes many giants of corporate social responsibility: Hamdi Ulukaya of Chobani, Emmanuel Faber of Danone, and Marc Benioff of Salesforce. Few people were surprised when, in June 2018, The B Team named a new chair: the now-retired CEO of Unilever, Paul Polman.

The members of The B Team try to lead by example, putting their own reputations on the line. Chobani not only pays its factory workers double the minimum wage, it implemented a profit-sharing plan so that employees collectively own 10 percent of the company. Danone has given its 100,000 employees a stake in company governance with its "One Person, One Voice, One Action" principle. The B Team is trying to gather together the pioneers who are pushing a different way forward for our corporations.

Rebuilding trust in our economy requires taking meaningful action. People won't be fooled time and again. Empty talk will be revealed for what it is, and it will breed even more cynicism when it's not backed up with action. To rebuild trust, our business leaders must make the first move. And they must make it count.

As Odysseus sailed his men home from the Trojan War, he tied himself to the mast to resist the Sirens' song. Throughout this book, we've seen business leaders build trust by tying themselves to the mast of social responsibility. This is what charters are all about: committing our corporations to a deeper purpose than profit and holding them accountable to that purpose. The Magna Carta—literally, "Great Charter"—involved King John of England pledging to his barons in 1215 that he would accept certain restrictions on his power in exchange for their allegiance. It was only by reluctantly tying himself to the mast that he could unify his kingdom. Sometimes our shackles set us free.

To rebuild trust, we must rebuild accountability: making owners accountable to corporations through stewardship codes; making corporations accountable through charters and standardized, mandatory ESG reporting; making university endowments accountable to students and faculty; making all our business leaders accountable to all their stakeholders. If fiduciary absolutism is fostered in distrust, citizen capitalism must be built on accountability.

We must continue to seek out authentic leadership and go work for those leaders, buy from them, and invest in their companies. We must use our economic power to build a better future—to forge, out of a distrustful economy and a fractured society, a capitalism defined not by our self-interest but by our citizenship.

EPILOGUE

A CAPITALIST REFORMATION

CATHEDRAL THINKING FOR OUR TIME

Jim Rogers had a saying: "If we are not at the table, we will be on the menu." It was a fitting stance for a man who fought tirelessly to combat climate change while also leading the nation's largest electric utility as the CEO of Duke Energy.

Rogers delivered double-digit returns to his shareholders over decades, an unheard-of performance in the lethargic world of utilities. And he did so while burning more coal than probably anyone else in the country. Nevertheless, he was best known for "using his perch as an industry leader to push the electric power sector to address the threat of climate change." His goal was to decarbonize the energy industry in this century.

Fred Krupp, who has led the Environmental Defense Fund for the last thirty years, said that Rogers "led by example" in laying the groundwork for comprehensive national climate legislation. He was the first prominent energy executive to recognize climate change and push for federal regulation to stem it. In 1990, he was the first CEO of a coal-based energy producer—at the time, he led the much smaller PSI Energy—to come out in favor of the Clean Air Act amendments. As one of his colleagues remembered it, the rest of the industry saw him as "this little pissant utility CEO" pushing the government to increase its regulation on all of them. Over the next couple decades, he would rise to become the largest pissant utility CEO, still pushing for increased environmental regulation.

In 2006, Rogers formed the US Climate Action Partnership to cut emissions by 80 percent by 2050. In 2001, he shocked fellow energy executives by supporting a federal cap on carbon dioxide. Between 2005 and 2018, Duke Energy invested $7 billion in wind and solar power and reduced its carbon emissions by 31 percent. After retiring, Rogers wrote a book about providing clean energy to the billion people on Earth without electricity.

Some saw Rogers as the height of hypocrisy—the most egregious example of those who have profited from dirty business turning around and attempting to cleanse their legacy. But Rogers didn't see his work as inconsistent with his advocacy. It was only his position within the energy industry that gave his advocacy real power.

A 2008 profile in the *New York Times*, titled "A Green Coal Baron?," described Rogers as "a genuine anomaly . . . he represents one of the country's biggest sources of greenhouse gases," yet he "is also one of the electricity industry's most vocal environmentalists." He forcefully advocated for greater regulation of greenhouse gases. Under his leadership, Duke cut the proportion of its energy coming from coal almost in half, from 58 percent in 2005 to 33 percent in 2017.

As the *Times* profile put it, "Rogers's environmentalism is practical, enthusiastic and intrigued by clean-tech innovations, not given to heartstring-tugging rhetoric about vanishing species or redwood trees." He accepted the reality of our current electrical grid but refused to accept a future in which the carbon we burn will destroy our climate.

For Rogers, combating climate change required "a long-term perspective, an appreciation of the complexity and scale of the problem, and a willingness to undertake investments today whose primary benefits will accrue to generations of the distant future." This is the central tenet around which we must rebuild our corporations: a commitment to the long term and a deeper purpose than profit.

His customers were a third residential, a third commercial, and a third industrial. Though electricity makes up a relatively small cost for households, it is the largest cost for many industrial customers. Rogers wanted to push for change that would not just appease environmentalists but also work for the twenty-five states that got a majority of their power from coal. He didn't want to increase costs so much for his industrial customers that they would up and leave for cheaper regions or cheaper countries. "He would want to balance all of that," remembered one colleague. "He was very much trying to find common ground, to take the long view."

When run by leaders like Rogers, utilities can approximate the long-term, stakeholder-oriented mindset we've seen so much of in this book. When an area's economy is doing better, its electricity use increases, and the utility makes more money. So Rogers's job was to promote economic development. Further, he thought not in years but decades. He made decisions that the next half dozen of his successors would have to live with, just as he lived with the decisions of his predecessors. Many of the nuclear plants built in the 1980s will still be operating in 2050.

Fiduciary absolutism has divided us against ourselves. It has told us that we must divorce our identities as owners from our identities as citizens and human beings. But to reform capitalism, we must be both moral and economic, idealistic and practical. Only by doing so can we restore the balance to our economy that fiduciary absolutism has undermined. It requires the mindset of pioneers such as Jim Rogers.

A SELF-FULFILLING DYSTOPIA

"Few trends," Milton Friedman wrote in 2002, "could so thoroughly undermine the very foundations of our free society as the acceptance by corporate officials of a social responsibility other than to make as much money for their stockholders as possible." Fiduciary absolutism is one hell of a drug.

Economists have run experiments on something they call the Dictator Game. In the Dictator Game, random participants are paired off, and then one of them is given $10 to split between them. Here's the catch: the other participant has zero say. Whatever the dictator decides is final. As in: game over, thanks for coming, pick up your cash on the way out. The economically rational thing would be for the dictator to take all $10.

Yet we don't. Even in the contrived, anonymous setting of the lab, more than 60 percent of dictators give money to the recipient; on average, they give a fifth of the pool. In fact, *only a third* of participants in economics games exhibit purely self-regarding behavior. Classical economics sees us each as an individual, atomistic utility maximizer: *Homo economicus.* But that's not who we are.

How else can we describe charitable giving in the United States? We donated $428 billion to charity in 2018. And although on a dollar basis 63 percent of individual donations are given by the top fifth of income earners, it is actually the lowest-income people who gave the most as a percentage of their income—some estimate they gave nearly triple the national percentage. They aren't giving to put their name onto a building or attend a black-tie benefit gala. They are giving because they want to give, because they feel a duty to give. *Homo economicus* doesn't exist any more than the invisible hand does. These are simplifying and in a way dehumanizing constructs.

The trouble is, they are constructs that end up shaping our world. In his book *The Moral Economy,* the economist Samuel Bowles documented the countless studies that have shown how economic incentives crowd out moral behavior. Strangers are less likely to help move a couch if they're offered payment than when they are asked to do so as a favor. When a day care center started charging parents who pick up their children late, more parents showed up late. Parents used to feel guilty for making teachers stay behind, but now they treated their own irresponsibility as just another market transaction. Even when the fee was removed, the tardiness continued because the social bond was severed. None of us is born *Homo economicus*, but our economy is good at making us so.

Against the cynics who paint us as rational, self-regarding utility maximizers, we have the reality: humans are social animals. We care about others for their own sake. Yet capitalism seems to throw these selfless instincts into jeopardy. "By propagating ideologically inspired amoral theories," wrote the late Sumantra Ghoshal, then a business professor at the London School of Economics, "business schools have actively freed their students from any sense of moral responsibility." As one economist put it, in economics "we label as 'irrational' not committing a crime when the expected benefit exceeds the expected punishment." Given the 372,000 bachelor's degrees and 187,000 master's degrees granted in business each year, this is terrifying.

Ghoshal recognized the costs incurred by teaching business students an ideology built on selfishness. As his colleagues decried corporate malfeasance like that of Enron, he noted, "Our theories and ideas have done much to strengthen the management practices that we are all now so loudly condemning." Indeed, there is one group that reliably plays the Dictator Game the selfishly maximizing way economists would predict: economics students. Teaching economics to students makes them more selfish, more likely to be tempted to corruption, and less concerned with the common good.

To abide by this view of human nature is not just to get us wrong but to make us wrong. The adage that "private vice makes for public virtue" fuels private vice while lessening public virtue; it's the self-defeating conclusion of a self-fulfilling premise. We can build a better economy, but it's going to require a higher aspiration than Milton Friedman offered.

Jonathan Sacks, former chief rabbi of the United Kingdom, once said:

> When everything that matters can be bought and sold, when commitments can be broken because they are no longer to our advantage, when shopping becomes salvation and advertising slogans our litany, when our worth is measured by how much we earn and spend, then the market is destroying the very virtues on which in the long run it depends.

Our society has suffered the reign of fiduciary absolutism for nearly fifty years. When will we recognize that our economy is capable of so much more?

We've seen throughout this book the business leaders and pioneers who have extended the frontier of responsible capitalism. But there are millions of unsung heroes at call centers and retail outlets, in voting booths and job fairs, who do so on a daily basis.

We've always had local business owners sponsoring Little League teams and employers who gave a little extra time off to employees struggling at home. In David Foster Wallace's description, we've always had the "low-wage clerk at the motor vehicle department," who, showing uncommon humanity, "helped your spouse resolve a horrific, infuriating, red-tape problem through some small act of bureaucratic kindness." In the end, there is no invisible hand. There's only us.

A CAPITALIST REFORMATION

The Wealth of Nations was not Adam Smith's first book. It was the sequel to an earlier work that he deemed much more important:

The Theory of Moral Sentiments. Smith was, at root, a philosopher. The field of economics did not even exist when he was writing. His description of how nations trade and become wealthy, of how markets operate, was in fact only a component of a larger argument about how we, as individuals, interact. The most important thing to Smith was our moral relations.

For Smith, economic efficiency was subservient to the larger goal of a nation of moral persons, citizens with a deep sense of obligation to one another. He started with the premise that man is a social animal: "How selfish soever man may be supposed, there are evidently some principles in his nature, which interest him in the fortune of others, and render their happiness necessary to him, though he derives nothing from it except the pleasure of seeing it."

Smith went on to enlist our morality in building a more just society, worrying that "we may often fulfil all the rules of justice by sitting still and doing nothing." It is not rules that can make us moral; morality can't be imposed on us. Smith didn't expect us to be angels: "To man is allotted a much humbler department, but one much more suitable to the weakness of his powers, and to the narrowness of his comprehension; the care of his own happiness, of that of his family, his friends, his country." Modern economists misread Smith by focusing only on the first part—the selfishness—and forgetting the rest.

"The wise and virtuous man," Smith wrote, "is at all times willing that his own private interest should be sacrificed to the public interest." This is the deeper foundation of capitalism today, a foundation that's hidden from view. "How can it be," asked Pope Francis in 2013, "that it is not a news item when an elderly homeless person dies of exposure, but it is news when the stock market loses two points?" Today we talk about financial markets as being driven alternately by fear and greed. For Smith, there was something more. We should add hope, purpose, a common project, and a shared belief.

In a sense, then, what we advocate for is not so much a break from capitalism as a return to its roots.

We have erred in believing that capitalism allows only for greed. We have built organizations that pursue narrow interests at any cost—within the law wherever possible, though sometimes not even then. Many of us have spent a majority of our lives serving these organizations. If capitalism is our church, we are in need of a reformation. Just as the Protestant Reformation was a return to the original text of the Bible, away from the extravagances of the Church that had coopted it, we are proposing a Capitalist Reformation that jettisons the baroque financialization of the economy and reinstills the latent moral character and social purpose at the root of all social organizations and all individuals.

Walk around any city or college campus, and you will notice something on many of the buildings (churches included): they are inscribed with the year they were built. It's not surprising that we put our names on things in the hope that, like Percy Bysshe Shelley's Ozymandias, our name will live on. But why the year?

As Wendell Berry put it, the world is not given to us by our fathers but borrowed from our children. Each generation wants to leave its mark, to tell for all time that it was the people of 1793 who built the Capitol, the people of 1930 who built the tallest building in the world. Above the speaker's rostrum in the House of Representatives, inscribed deep into the wood, are the words of Daniel Webster: "Let us develope the resources of our land, call forth its powers, build up its institutions, promote all its great interests and see whether we also in our day and generation may not perform something worthy to be remembered."

Shortly before Christmas in 2018, Jim Rogers was visiting family in Louisville, Kentucky, when he suddenly died, at the age of seventy-one. He had previously retired from his post as the renegade environmentalist leading Duke Energy. In 2010, he had testified in front of the Senate about reducing our economy's addiction to short-term profits. "If you go to Europe and you look at the great cathedrals," he told the Senate, "you recognize that they were built over 100 years, most of them." He continued:

> You are looking at three to four generations to build a cathedral. And so the people that worked on the foundations never saw the walls or the stained glass windows. Those that worked on the stained glass windows never saw the spires. The architect never saw it finished. And yet every generation . . . committed their time, their energy, their passion to getting it done.

He called this cathedral thinking. The generations that built the cathedrals would never see the centuries of baptisms and weddings, worship and funerals for which they laid the foundation. But they worked anyway, because they had faith in the future. "Their vision was as clear as the future ringing of that bell on a cold winter morning," he said. "Their vision was of something bigger than themselves."

Jim Rogers saw combating climate change as today's cathedral project. Making capitalism reflect our values is, too.

In 1850, the tallest building in New York City was Trinity Church, rising 281 feet above Broadway. At the time, none of the tallest buildings in the world was commercial. Today, the tallest building in New York City is One World Trade Center. In almost every city in the United States, the tallest building is not a church or a government building but an office tower. President Calvin Coolidge said, "The man who builds a factory builds a temple; the man who works there worships there." That sounds sad. It doesn't have to be.

In the end, corporations are just social organizations we create to attain a common end. We all want to be a part of something bigger than ourselves. That's all a corporation is: something bigger than the sum of its parts. Whether it's good or bad that our lives are dominated by corporations depends on the sorts of corporations we choose to build. "Make no little plans," said the Chicago architect Daniel Burnham. "They have no magic to stir men's blood and probably themselves will not be realized." Fiduciary absolutism—maximizing short-term profits for shareholder value—has rarely stirred much blood.

For those who are building corporations today, the most important question will be what purposes can rise to Burnham's challenge—what purposes can inspire our dedication and are worthy of our efforts.

"We must take the same approach as those cathedral builders took centuries ago," Rogers said. "We can't change the world in one day, one week, one month, one year or one decade. We must build on our commitment over time, and have faith that our work will eventually achieve our highest aspirations."

It's our economy, our society. What are we going to do with it? We can build something that would make our forebears proud and posterity forever grateful. We are accountable to no one but one another, to no one but those to whom we will leave this world. And so we face our final charge: *whether we also in our day and generation may not perform something worthy to be remembered.*

ACKNOWLEDGMENTS

In this book, we have tried to share the stories and insights of pioneers—those CEOs, start-up founders, investors, academics, social activists, and others who are shaping the future of capitalism. Thank you to Frank Britt, Ilana Cohen, Sir Ronald Cohen, Tony Davis, Chad Dickerson, Guy Dixon, Curt Ellis, Jed Emerson, Galen Hall, Ned Hall, Chris James, Andrew Kassoy, Chris Klebba, Rich Klein, Clay Lambert, John Lauve, Patrick McKiernan, Kate Murtagh, Jon Olinto, Ram Palaniappan, Charles Penner, Cliff Robbins, Barry Rosenstein, Dave Scanzoni, Caleb Schwartz, Josh Silverman, Céline Soubranne, former chief justice Leo Strine, Halla Tómasdóttir, Jeff Ubben, Ed Walsh, and Tom Williams.

Over the last two years, we benefited from the feedback of friends, colleagues, professors, and experts in business, finance, sustainability, public policy, and beyond. This book is far stronger because of their many critiques. A special thank-you to Ryan Beck, R. B. Brenner, Paul Brest, Tim Brown, Stephen Cameron, Camille Canon, Steve Davis, Larry Kramer, Glenn Kramon, Jonathan Levin, Josh Mitrani, Amit Seru, Jacob Stern, Gal Treger, Geoff Tuff, Richard Valdmanis, Robert Valdmanis, Thor Valdmanis, Naicheng Wangyu, Owen Wurzbacher, and Chris Zook. We owe much of this project to Governor Deval Patrick, who made impact-investing converts of us both and continues to inspire us to aim higher.

An elite crew of professionals made this book a reality. Thank you to Lynn Anderson, Janet Byrne, Jessica Chao, Cassidy Donahue, Evelyn Duffy, and Penelope Lin for the critical research, editing, and fact checking they provided. Thank you to our editor at HarperCollins, Hollis Heimbouch, for believing in this book and this vision. Thank you to Martina O'Sullivan, our editor at Penguin, for helping to bring this book to an international readership. Thank you to our wonderful agent, Alice Martell, whose early support and boundless enthusiasm have propelled us forward. And thank you to the many friends who have helped us navigate this new world: Jordan Blashek, Sofia Groopman, Chris Haugh, Mike Lewis, and Diego Nunez.

From Michael: I want to thank Zoe, my partner through every step and stumble; my mother, the most naturally gifted storyteller I know; and my father, whom we lost too soon—but not before he showed us what it means to be good.

From Warren: I want to thank my wife, Kristin, whose love, advice, and support make all things possible; my three brothers, Rich, Rob, and Thor, whose thirst for debate turns every family dinner into a public-policy session; my mother and father, who have shown me what it means to care about others; and my children, Lizzie, Ian, Anna, and Teddy, who keep me smiling and give me hope for the future.

NOTES

INTRODUCTION: SAVING CAPITALISM FROM ITSELF

1 tenfold increase in applications: Christian Nesheim, "New Zealand's Investor Visa Raises NZ$27 Million a Week, Rejects 1 in 3 Applicants," *Investment Migration Insider*, May 6, 2019, https://www.imidaily.com/asia-pacific/new-zealands-investor-visa-raises-nz27-million-a-week-rejects-1-in-3-applicants/.

1 "Saying you're 'buying a house'": Mark O'Connell, "Why Silicon Valley Billionaires Are Prepping for the Apocalypse in New Zealand," *The Guardian*, February 15, 2008, https://www.theguardian.com/news/2018/feb/15/why-silicon-valley-billionaires-are-prepping-for-the-apocalypse-in-new-zealand.

2 "I feel better": See https://survivalcondo.com/overview/.

2 "I know of no country": Quoted in Alan Greenspan and Adrian Wooldridge, *Capitalism in America: A History* (New York: Penguin, 2018), 9, 43.

2 America is a nation: As opposed to America the landmass, which was obviously founded by Native Americans.

2 Half the passengers: Bhu Srinivasan, *Americana: A 400-Year History of American Capitalism* (New York: Penguin, 2017), 12.

2 For those willing: Srinivasan, *Americana*, 9.

2 "with the risks borne": Srinivasan, *Americana*, 42–43.

2 85 percent of Virginia's settlers: Srinivasan, *Americana*, 18.

3 "This is a government": Quoted in Greenspan and Wooldridge, *Capitalism in America*, 167.

3 the gap in life expectancy: "Chronicle Reports on Lifespan Disparities in Boston," American Heart Association—Massachusetts, July 1, 2016, https://massachusetts.heart.org/2016/07/01/chronicle-explores-lifespan-disparities-boston/.

3 the largest fifty-three metro areas: Mark Muro and Jacob Whiton, "Geographic Gaps Are Widening While U.S. Economic Growth Increases," Brookings Institution, January 23, 2018, https://www.brookings.edu/blog/the-avenue/2018/01/22/uneven-growth/.

3 Half of rural counties: Kim Parker et al., "Demographic and Economic Trends in Urban, Suburban and Rural Communities," Pew Research Center, May 22, 2018, https://www.pewsocialtrends.org/2018/05/22/demographic-and-economic-trends-in-urban-suburban-and-rural-communities/.

3 children will "be better off": Bruce Stokes, "2. Expectations for the future," Pew Research Center, September 18, 2018, https://www.pewresearch.org/global/2018/09/18/expectations-for-the-future/.

3 half of Americans: Tony Schwartz and Christine Porath, "Why You Hate Work," *New York Times*, May 30, 2014, https://www.nytimes.com/2014/06/01/opinion/sunday/why-you-hate-work.html?_r=1. For more on lack of meaning at work, see David Graeber, *Bullshit Jobs* (New York: Simon & Schuster, 2018), and Charles Duhigg, "America's Professional Elite: Wealthy, Successful and Miserable," *New York Times Magazine*, February 21, 2019, https://www.nytimes.com/interactive/2019/02/21/magazine/elite-professionals-jobs-happiness.html.

4 one 2020 study: Mark John, "Capitalism Seen Doing 'More Harm Than Good' in Global Survey," Reuters, January 20, 2020, https://www.reuters.com/article/us-davos-meeting-trust/capitalism-seen-doing-more-harm-than-good-in-global-survey-idUSKBN1ZJ0CW.

4 view capitalism favorably: Lydia Saad, "Socialism as Popular as Capitalism Among Young Adults in U.S.," Gallup, November 25, 2019, https://news.gallup.com/poll/268766/socialism-popular-capitalism-among-young-adults.aspx.

4 prefer living in a socialist country: "'Axios on HBO' Poll: 55% of Women Prefer Socialism," Axios, June 9, 2019, https://www.axios.com/axios-hbo-poll-55-percent-women-prefer-socialism-f70bf87e-34fd-4b63-b1f6-2f2b6900f634.html.

5 higher than ever: "Corporate Profits Are at an All Time High," Center on Budget and Policy Priorities, n.d., https://www.cbpp.org/corporate-profits-are-at-an-all-time-high.

5 "Hegel predicted": John Micklethwait and Adrian Wooldridge, *The Company: A Short History of a Revolutionary Idea* (New York: Modern Library, 2003), xiv–xv.

6 You may have woken: "Fortune 500 2018," *Fortune*, May 21, 2018, https://fortune.com/fortune500/2018/.

6 "Commerce defies every wind": George Bancroft, "Bancroft Quotes," in *Forty Thousand Quotations: Prose and Poetical*, comp. C. N. Douglas (New York: Halcyon House, 1917), https://www.bartleby.com/348/authors/36.html.

7 Nearly nine in ten corporations: Governance and Accountability Institute, "Flash Report: 86% of S&P 500 Index® Companies Publish Sustainability/Responsibility Reports in 2018," May 16, 2019, https://www.ga-institute.com/press-releases/article/flash-report-86-of-sp-500-indexR-companies-publish-sustainability-responsibility-reports-in-20.html.

7 institutions representing $11 trillion: Yossi Cadan, Ahmed Mokgopo, and Clara Vondrich, *$11 Trillion and Counting: New Goals for a Fossil-Free World*, 350.org, September 9, 2019, https://financingthefuture.platform350.org/wp-content/uploads/sites/60/2019/09/FF_11Trillion-WEB.pdf.

7 Impact investing has become: Abhilash Mudaliar and Hannah Dithrich, "Sizing the Impact Investing Market," Global Impact Investing Network, April 1, 2019, https://thegiin.org/assets/Sizing%20the%20Impact%20Investing%20Market_webfile.pdf.

7 $80 trillion is now covered: Principles for Responsible Investment, brochure, January 28, 2019, https://www.unpri.org/download?ac=6303.

7 two-thirds of Americans: Luigi Zingales, "Does Finance Benefit Society?," presidential address, American Finance Association Annual Meeting, Boston, MA, January 4, 2015, https://faculty

.chicagobooth.edu/luigi.zingales/papers/research/finance .pdf, 2.

7 Our thousand largest corporations: Ronald J. Gilson and Jeffrey N. Gordon, "The Agency Costs of Agency Capitalism: Activist Investors and the Revaluation of Governance Rights," *Columbia Law Review* 113 (2013): 874.

7 "Our economy is dominated": David Ciepley, "Beyond Public and Private: Toward a Political Theory of the Corporation," *American Political Science Review* 107, no. 1 (February 2013): 147.

7 CEOs' pay has increased: Lawrence Mishel and Julia Wolfe, "CEO Compensation Has Grown 940% Since 1978," Economic Policy Institute, August 14, 2019, https://www.epi.org/publication/ceo -compensation-2018/.

8 buying back $4 trillion: Chuck Schumer and Bernie Sanders, "Limit Corporate Stock Buybacks," *New York Times*, February 3, 2019, https://www.nytimes.com/2019/02/03/opinion/chuck -schumer-bernie-sanders.html; William Lazonick, "Profits Without Prosperity," *Harvard Business Review*, September 2014, https:// hbr.org/2014/09/profits-without-prosperity.

9 Icelandic rock band Hatari: Rob Holley, "Iceland's Hatari: 'At Eurovision, We're the Pink Elephant in the Room,'" *The Independent*, May 18, 2019, https://www.independent.co.uk/arts-entertainment /music/features/hatari-interview-eurovision-2019-iceland-song -israel-tel-aviv-boycott-a8818626.html.

9 "Our economic system": Naomi Klein, *This Changes Everything: Capitalism vs. the Climate* (New York: Simon & Schuster, 2014), 21.

9 Only one in four people: Miguel Padró, "Unrealized Potential: Misconceptions about Corporate Purpose and New Opportunities for Business Education," The Aspen Institute, May 28, 2014, https://assets.aspeninstitute.org/content/uploads/files/content /docs/pubs/Aspen%20BSP%20Unrealized%20Potential%20 May2014v.2.pdf?_ga=2.40107429.603537148.1576800305 –420537432.1576800305, 2.

10 "They pretend to pay us": Referencing the joke, see Paul Collier, *The Future of Capitalism: Facing the New Anxieties* (New York: HarperCollins, 2018), 18.

10 "This had very good success": Quoted in Srinivasan, *Americana*, 15–16.

10 During the three millennia: J. Bradford De Long, "Estimates of World GDP, One Million B.C.–Present," 1998, https://delong.typepad.com/print/20061012_LRWGDP.pdf.

10 skyrocketed by more than: De Long, "Estimates of World GDP, One Million B.C.–Present."

10 Just since 1990: World Bank, "Poverty," https://www.worldbank.org/en/topic/poverty/overview.

10 "I weigh my words": Quoted in Greenspan and Wooldridge, *Capitalism in America*, 133.

10 "We can see how": Darren Walker, "In Defense of Nuance," Ford Foundation, September 19, 2019, https://mailchi.mp/d3497b214100/in-defense-of-nuance?e=d42d0e45ea.

11 these companies represent: See "550: Fortune 500," https://fortune.com/fortune500/.

11 Gilded Age inequality: For parallels to the Gilded Age as well as the pitfalls of such comparisons, see David Huyssen, "We Won't Get Out of the Second Gilded Age the Way We Got Out of the First," Vox, April 1, 2019, https://www.vox.com/first-person/2019/4/1/18286084/gilded-age-income-inequality-robber-baron.

11 Magie's original game: Mary Pilon, "Monopoly's Inventor: The Progressive Who Wouldn't Pass 'Go,'" *New York Times*, February 13, 2015, https://www.nytimes.com/2015/02/15/business/behind-monopoly-an-inventor-who-didnt-pass-go.html; Mary Pilon, "Monopoly Was Designed to Teach the 99% About Economic Inequality," *Smithsonian*, January 2015, https://www.smithsonianmag.com/arts-culture/monopoly-was-designed-teach-99-about-income-inequality-180953630/. Both stories draw on Pilon's book *The Monopolists: Obsession, Fury, and Scandal Behind the World's Favorite Board Game* (New York: Bloomsbury, 2015).

12 It's like the joke: Cohen is a ghost (@skullmandible), "is there anything more capitalist than a peanut with a top hat, cane, and monocle selling you other peanuts to eat," Twitter, August 29, 2013, 11:42 a.m., in "Well ya got me there," post by

u/EnterMyCranium, r/LateStageCapitalism, Reddit, February 19, 2019, https://www.reddit.com/r/LateStageCapitalism/comments/as86l8/well_ya_got_me_there/.

12 "a steam-powered locomotive": Jill Lepore, *These Truths: A History of the United States* (New York: W. W. Norton, 2018), 198.

12 "Now at length": Benjamin Franklin, September 17, 1787, speech, Madison Debates, The Avalon Project Documents in Law, History and Diplomacy, Lillian Goldman Law Library, Yale Law School, https://avalon.law.yale.edu/18th_century/debates_917.asp.

12 "We have been dosing": Lynn A. Stout, "The Problem of Corporate Purpose," *Issues in Governance Studies* 48 (June 2012), https://www.brookings.edu/wp-content/uploads/2016/06/Stout_Corporate-Issues.pdf, 4.

CHAPTER 1: CHALK ONE UP FOR THE GOOD GUYS

13 "I'm doing what I'm doing": Guy Dixon, interview by Michael O'Leary, November 21, 2019. Unless noted otherwise, all further remarks from Dixon are from this interview.

13 the average holding period: World Bank, "Stock Market Turnover Ratio (Value Traded/Capitalization) for United States (DDEM01USA156NWDB)," Federal Reserve Bank of St. Louis, February 19, 2020, https://fred.stlouisfed.org/series/DDEM01USA156NWDB.

14 A third of new CEOs: David F. Larcker and Brian Tayan, "Internal vs. External CEOs: Research Spotlight," Stanford Graduate School of Business, October 18, 2016, https://www.gsb.stanford.edu/sites/gsb/files/publication-pdf/cgri-research-spotlight-08-internal-versus-external-ceos.pdf.

14 the median job tenure: US Bureau of Labor Statistics, "Employee Tenure in 2018," September 20, 2018, https://www.bls.gov/news.release/tenure.nr0.htm.

14 the broader economy was firing: Center on Budget and Policy Priorities, "Chart Book: The Legacy of the Great Recession," June 6, 2019, https://www.cbpp.org/research/economy/chart-book-the-legacy-of-the-great-recession.

15 In their lawsuit: Laurence V. Parker, Jr., "Corporate and Business

Law," *University of Richmond Law Review* 48, no. 39 (November 2013): 39–62, https://lawreview.richmond.edu/files/2013/11/Parker-481.pdf.

16 "battle for the soul of capitalism": William W. George and Amram Migdal, "Battle for the Soul of Capitalism: Unilever and the Kraft Heinz Takeover Bid (A)," Harvard Business School, May 30, 2017, https://store.hbr.org/product/battle-for-the-soul-of-capitalism-unilever-and-the-kraft-heinz-takeover-bid-a/317127.

16 Andrew Carnegie's "The Gospel of Wealth": "Wealth," *North American Review* 148, no. 391 (June 1889): 653–65. Published later in book form as *The Gospel of Wealth, and Other Timely Essays* (New York: The Century Company, 1900). Available at Carnegie Corporation of New York, https://www.carnegie.org/about/our-history/gospelofwealth/.

16 a taped interview: Unilever and *HuffPost*, "Talk to Me," featuring Paul Polman and Sebastian Polman, https://www.youtube.com/watch?v=nARmfhXhCQg.

17 "near-death" experience: Scheherazade Daneshkhu, "Pressure Is on Unilever to Meet Investor Expectations," *Financial Times*, March 14, 2017, https://www.ft.com/content/e1e35e36-08cd-11e7-ac5a-903b21361b43.

17 "there are so many": Terry Slavin, "Paul Polman: 'I Feel Like It's My First Day at Unilever, and There's a Lot to Do,'" Ethical Corporation/Reuters Events, May 1, 2018, http://www.ethicalcorp.com/paul-polman-i-feel-its-my-first-day-unilever-and-there-lot-do.

18 "[He] would no more guzzle down a Magnum": James Ashton, "Why a Giant Food Fight Is Breaking Out over Unilever," *Telegraph*, February 18, 2017, https://www.telegraph.co.uk/business/2017/02/18/giant-food-fight-breaking-unilever/.

18 "our greatest resource": "Alex Behring Reveals His Investment Philosophy," *Financial Times*, May 7, 2017, https://www.ft.com/content/850c39a2-330d-11e7-bce4-9023f8c0fd2e.

18 shuttered seven North American plants: Julie Creswell and David Yaffe-Bellany, "When Mac & Cheese and Ketchup Don't Mix: The Kraft Heinz Merger Falters," *New York Times*, September 24, 2019, https://www.nytimes.com/2019/09/24/business/kraft-heinz-food-3g-capital-management.html; Teresa F. Lindeman, "Kraft

Heinz to Close 7 North American Plants, Cut 2,600 Jobs," *Pittsburgh Post-Gazette*, November 4, 2015, https://www.post-gazette.com/business/pittsburgh-company-news/2015/11/04/Kraft-Heinz-to-close-7-North-American-plants-cut-2-600-jobs/stories/201511040186; Julie Jargon, "Kraft to Slash 2,600 More Jobs, Close Seven Plants," *Wall Street Journal*, November 4, 2015, https://www.wsj.com/articles/kraft-to-slash-2-600-more-jobs-close-seven-plants-1446668634.

19 closed a factory in Keynsham: "Cadbury Plant in Keynsham Put Up for Sale by Kraft," BBC News, March 23, 2011, https://www.bbc.com/news/uk-england-bristol-12818328.

19 600 jobs, now gone: "Cadbury's Somerdale Plant in Keynsham Sold to Developer," BBC News, January 4, 2012, https://www.bbc.com/news/uk-england-bristol-16410771.

19 "utterly despicable": "Cadbury Factory Closure by Kraft 'Despicable,'" BBC News, February 10, 2010, http://news.bbc.co.uk/2/hi/business/8507780.stm.

19 laid off more than ten thousand people: Brigid Sweeney, "Kraft Heinz CEO: It's All About Long-Term Perspective, Not Ruthless Cost-Cutting," *Crain's Chicago Business*, September 12 2017.

19 quickly amended its packaging: Rebecca Smithers and Mark Sweney, "Heinz Pulls Ads for 'Healthy' Baby Biscuits After Complaint to ASA," *The Guardian*, September 8, 2015, https://www.theguardian.com/lifeandstyle/2015/sep/09/heinz-pulls-ads-asa-healthy-baby-biscuits.

19 in the fiftieth percentile: "The Kraft Heinz Company CSR/ESG Ranking," CSRHub, last modified October 2019, https://www.csrhub.com/CSR_and_sustainability_information/The-Kraft-Heinz-Company.

19 Unilever ranks near the top: "Unilever NV CSR/ESG Ranking," CSRHub, last modified October 2019, https://www.csrhub.com/CSR_and_sustainability_information/Unilever-NV.

19 "a clash": "Paul Polman: In His Own Words," *Financial Times*, December 3, 2017, https://www.ft.com/content/2209d63a-d6ae-11e7-8c9a-d9c0a5c8d5c9.

20 "Unilever does not see the basis": David Gelles, "Why the Kraft Heinz Bid for Unilever Could Make an Odd Match," *New York*

Times, February 17, 2017, https://www.nytimes.com/2017/02/17/business/dealbook/why-the-kraft-heinz-bid-for-unilever-could-make-an-odd-match.html.

20 "We look forward": Liz Moyer, "The 'Warren Buffett of Brazil' Behind the Bid for Unilever," *New York Times*, February 17, 2017, https://www.nytimes.com/2017/02/17/business/dealbook/the-warren-buffett-of-brazil-behind-the-offer-for-unilever.html.

20 "make cleanliness commonplace": "Unilever History, 1871–2019," https://www.unileverusa.com/about/who-we-are/our-history/.

20 By the end of the century: James O'Toole, *The Enlightened Capitalists: Cautionary Tales of Business Pioneers Who Tried to Do Well by Doing Good* (New York: Harper Business, 2019), 51.

21 "a high standard of housing": "Unilever History, 1871–2019," https://www.unileverusa.com/about/who-we-are/our-history/.

21 "My job is not": Bill George, *Discover Your True North: Becoming an Authentic Leader* (Hoboken: Wiley, 2015), 244.

22 "Harvard Business School professor": Alan Murray, "America's CEOs Seek a New Purpose for the Corporation," *Fortune*, August 19, 2019, https://fortune.com/longform/business-roundtable-ceos-corporations-purpose.

22 "Every time I have made": Yvon Chouinard, "The Way I Work: Yvon Chouinard, Patagonia," *Inc.*, March 12, 2013, https://www.inc.com/magazine/201303/liz-welch/the-way-i-work-yvon-chouinard-patagonia.html.

22 Mackey's book: The book is *Conscious Capitalism: Liberating the Heroic Spirit of Business* (Boston: Harvard Business Review Press, 2013).

22 founded in 2006: B Lab, "How Did the B Corp Movement Start?," https://bcorporation.net/faq-item/how-did-b-corp-movement-start.

22 certified by the nonprofit B Lab: B Lab, "About B Corps," Certified B Corporation, July 31, 2018, https://bcorporation.net/about-b-corps.

22 "are accelerating": Ibid.

22 There are now more than 3,000: B Lab/Sistema B Global Network, letter, December 11, 2019, Madrid, https://bcorporation.net/news/500-b-corps-commit-net-zero-2030.

22 most B Corps are: Miriam Landman, "Best B Corporations, 2018," The Green Spotlight, July 30, 2018, https://www.thegreenspotlight.com/2018/07/b-corporations-2018/.

22 "The theory of change": Andrew Kassoy, interview by Michael O'Leary, February 5, 2020. Unless otherwise noted, all further remarks from Kassoy are from this interview.

23 of the 3,700 or so: Jason M. Thomas, "Where Have All the Public Companies Gone?," *Wall Street Journal*, November 16, 2017, https://www.wsj.com/articles/where-have-all-the-public-companies-gone-1510869125.

23 only three are B Corps: B Lab, "Are Any B Corps Publicly Traded?," Certified B Corporation, July 31, 2018, https://bcorporation.net/faq-item/are-any-b-corps-publicly-traded.

23 largest certified B Corp: Leon Kaye, "Danone North America Is Now the Largest B Corp on Earth," Triple Pundit, April 13, 2018, https://www.triplepundit.com/story/2018/danone-north-america-now-largest-b-corp-earth/12846.

23 plans to become a B Corp: Sapna Maheshwari, "Gap Plans to Spin Off Old Navy After a Dismal Year," *New York Times*, February 28, 2019, https://www.nytimes.com/2019/02/28/business/gap-old-navy-spinoff.html.

23 achieved the year before: Gap Inc., "Athleta Earns a B Corp Certification," press release, March 20, 2018, https://www.gapinc.com/content/gapinc/html/media/pressrelease/2018/med_pr_032018_Athleta_BCorp.html.

23 The Body Shop also: The Body Shop, "The Body Shop issues call to action as it announces B Corp Certification," press release, September 23, 2019, https://markets.businessinsider.com/news/stocks/the-body-shop-issues-call-to-action-as-it-announces-b-corp-certification-1028545622.

23 business judgment rule: Leo E. Strine Jr., "The Delaware Way: How We Do Corporate Law and Some of the New Challenges We (and Europe) Face," *Delaware Journal of Corporate Law* 30 (2005): 681.

24 "Business is here to serve society": Paul Polman, "Business, Society, and the Future of Capitalism," *McKinsey Quarterly*, May 2014,

https://www.mckinsey.com/business-functions/sustainability/our-insights/business-society-and-the-future-of-capitalism.

24 "Why should the citizens": Andrew Edgecliffe-Johnson, "Beyond the Bottom Line: Should Business Put Purpose Before Profit?," *Financial Times*, January 4, 2019, https://www.ft.com/content/a84647f8-0d0b-11e9-a3aa-118c761d2745.

24 operating profit margins: The Kraft Heinz Company, Q4 2016 and Full Year Update, 14, http://ir.kraftheinzcompany.com/static-files/cbf6a943-4d7f-47e0-a506-c9f5bfb65d48; Unilever, Annual Report and Accounts 2016, 23, https://www.unilever.com/Images/unilever-annual-report-and-accounts-2016_tcm244-498744_en.pdf.

24 "This is my kind of deal": "Warren Buffett's Berkshire in $23 Billion Deal to Buy Heinz," NBC, February 14, 2013, https://www.nbcnews.com/business/business-news/warren-buffetts-berkshire-23-billion-deal-buy-heinz-flna1C8369091.

24 "could not be better partners": Warren E. Buffett, Berkshire Hathaway Inc. 2015 letter to shareholders, February 27, 2016, 6, https://www.berkshirehathaway.com/letters/2015ltr.pdf.

25 "to be more productive": Andrew Ross Sorkin, "Warren Buffett's Case for Capitalism," *New York Times*, May 5, 2019, https://www.nytimes.com/2019/05/05/business/warren-buffett-capitalism.html.

25 For Buffett, that problem: Chip Cutter, "Warren Buffett Says AI Will Lead to Fewer Jobs, Warning Future Could Be 'Enormously Disruptive,'" LinkedIn, May 6, 2017, https://www.linkedin.com/pulse/warren-buffett-predicts-significantly-less-employment-chip-cutter/.

25 Buffett sees the growing: Sorkin, "Warren Buffett's Case."

25 "I would run screaming": David Bank, "Bill Gates Is Putting His Own Money into a Small Impact-Investing Fund Focused on India," Quartz, November 16, 2014, https://qz.com/297097/bill-gates-unitus-seee-fund-impact-investing-fund-focused-on-india/.

25 "I like our way": Sorkin, "Warren Buffett's Case."

25 a compelling case: Milton Friedman, "The Social Responsibility of Business Is to Increase Its Profits," *New York Times Magazine*,

September 13, 1970, https://timesmachine.nytimes.com/timesmachine/1970/09/13/223535562.html?pageNumber=375.

26 working for the shareholders: Saying that shareholders "own" a corporation reflects the way laymen use the word but won't get us published in any academic journals. Technically, a corporation owns itself as a legal person. Shareholders do not own corporations the way they might own their cars. But they have ultimate control, so we'll stick with the term "ownership" even if, technically, shareholders just own shares that, among other rights, allow them to elect a board, which hires a CEO, who runs the company. If this all feels needlessly hairsplitting, law school might not be for you.

27 "My wealth has come": Warren Buffett, "My Philanthropic Pledge," The Giving Pledge, June 16, 2010, https://givingpledge.org/Pledger.aspx?id=177.

27 it can be traced: Anand Giridharadas, *Winners Take All: The Elite Charade of Changing the World* (New York: Knopf, 2018). Giridharadas pointed out that charity itself can be a form of corrupting power.

27 U.S. Steel employed 250,000: Alan Greenspan and Adrian Wooldridge, *Capitalism in America: A History* (New York: Penguin, 2018), 123.

27 sixteen people died: "The Strike at Homestead Mill," NPR, https://www.pbs.org/wgbh/americanexperience/features/carnegie-strike-homestead-mill/.

27 he amassed a fortune: "Andrew Carnegie: Pittsburgh Pirate," *The Economist,* January 30, 2003, https://www.economist.com/books-and-arts/2003/01/30/pittsburgh-pirate.

27 "The man who dies thus rich": Andrew Carnegie, "The Gospel of Wealth," June 1889, Carnegie Corporation of New York, https://www.carnegie.org/about/our-history/gospelofwealth/.

27 "millionaire's unworthy life": Ibid.

27 The Philanthropy Roundtable declared: Leslie Lenkowsky, "Andrew Carnegie," Philanthropy Hall of Fame, March 23, 2013, https://www.philanthropyroundtable.org/almanac/people/hall-of-fame/detail/andrew-carnegie.

28 He endowed 2,811 libraries: "Andrew Carnegie: Pittsburgh Pirate."

28 Fatal accidents in steel mills: Ibid.

28 "acknowledgement of the deep debt": Bhu Srinivasan, *Americana: A 400-Year History of American Capitalism* (New York: Penguin, 2017), 249.

28 "The Gospel of Wealth" allowed him: Mark Twain once quipped, "I wanted to see some more of Andrew Carnegie, who is always a subject of intense interest for me. I like him; I am ashamed of him; and it is a delight to me to be where he is if he has new material on which to work his vanities." Mark Twain, *Autobiography of Mark Twain*, vol. 3, ed. Benjamin Griffin and Harriet Elinor Smith (Berkeley: University of California Press, 2015), 189.

28 Americans have consistently donated: Benjamin Soskis, "Giving Numbers: Reflections on Why, What, and How We Are Counting," *Nonprofit Quarterly* (Fall 2017: "The Changing Skyline of U.S. Giving"), https://nonprofitquarterly.org/giving-numbers-reflections-counting/.

28 An informal poll: Survey conducted by the authors, November 15, 2018.

29 "There is something within all of us": Martin Luther King, Jr., "Loving Your Enemies," sermon, November 17, 1957[?], https://kinginstitute.stanford.edu/king-papers/documents/loving-your-enemies-sermon-delivered-dexter-avenue-baptist-church.

29 "There is that persistent schizophrenia": Martin Luther King, Jr., "Love in Action," in *The Papers of Martin Luther King Jr.*, vol. 6, *Advocate of the Social Gospel, September 1948–March 1963,* edited by Clayborne Carson et al. (Berkeley and Los Angeles: University of California Press, 2007), https://kinginstitute.stanford.edu/king-papers/documents/draft-chapter-iv-love-action.

30 "And I had this thought": Chris James, interview by Michael O'Leary, December 4, 2019. Unless otherwise noted, all further remarks from James are from this interview.

30 "If Andrew Carnegie had employed": Quoted in O'Toole, *The Enlightened Capitalists*, xxxiii.

31 renamed in his honor: Robin Pogrebin, "A $100 Million Donation to the N.Y. Public Library," *New York Times,* March 11, 2008, https://www.nytimes.com/2008/03/11/arts/design/11expa.html.

31 "what you *don't* pay for": Amie Tsang, "5 Pieces of Advice from

Jack Bogle," *New York Times,* January 17, 2019, https://www.nytimes.com/2019/01/17/business/mutfund/john-bogle-vanguard-investment-advice.html.

31 he saves investors: William Baldwin, "Jack Bogle Is Gone, but He's Still Saving Investors $100 Billion a Year," *Forbes,* January 16, 2019, https://www.forbes.com/sites/baldwin/2019/01/16/jack-bogle-is-gone-but-hes-still-saving-investors-100-billion-a-year/#10214bf3795c.

32 40 percent of millennials: Adele Peters, "Most Millennials Would Take a Pay Cut to Work at an Environmentally Responsible Company," *Fast Company,* February 14, 2019, https://www.fastcompany.com/90306556/most-millennials-would-take-a-pay-cut-to-work-at-a-sustainable-company.

32 preferred sustainable brands: Dan Berthiaume, "Survey: Generations Differ on Importance of Sustainability," Chain Store Age, January 20, 2020, https://chainstoreage.com/survey-generations-differ-importance-sustainability.

33 the rational response is hypocrisy: Charles H. Cho, Matias Laine, Robin W. Roberts, and Michelle Rodrigue, "Organized Hypocrisy, Organizational Façades, and Sustainability Reporting," *Accounting, Organizations and Society* 40 (January 2015): 78–94.

34 "remove the temptation": Polman, "Business, Society, and the Future of Capitalism."

35 offering $50: Michael J. de la Merced and Chad Bray, "Kraft Heinz Offers to Buy Unilever in $143 Billion Deal," *New York Times,* February 17, 2017, https://www.nytimes.com/2017/02/17/business/dealbook/kraft-heinz-unilever-deal-.html.

35 "Unilever and Kraft Heinz hereby announce": Antoine Gara, "Kraft Heinz Withdraws Its $143 Billion Bid for Unilever," *Forbes,* February 19, 2017, https://www.forbes.com/sites/antoinegara/2017/02/19/kraft-heinz-withdraws-its-143-billion-bid-for-unilever/#2dabb2a94063.

35 "to avoid a potentially dirty": Arash Massoudi and James Fontanella-Khan, "The $143bn Flop: How Warren Buffett and 3G Lost Unilever," *Financial Times,* February 21, 2017, https://www.ft.com/content/d846766e-f81b-11e6-bd4e-68d53499ed71.

35 "I've been here before": Interview with hedge fund manager, October 2019.

36 "the market is a weighing machine": Warren E. Buffett, Berkshire Hathaway Inc. 1993 letter to shareholders, March 1, 1994, https://www.berkshirehathaway.com/letters/1993.html. Also see "'In the Short-Run, the Market Is a Voting Machine, but in the Long-Run, the Market Is a Weighing Machine': Benjamin Graham? Warren Buffett? Ronald A. McEachern? Ben Bidwell? John C. Bogle? Apocryphal?," Quote Investigator, updated January 9, 2020, accessed January 13, 2020, https://quoteinvestigator.com/2020/01/09/market/.

36 Unilever's share price has grown: Unilever historical share price, February 16, 2018–February 18, 2020, retrieved from Unilever, https://www.unilever.com/investor-relations/shareholder-centre/share-prices/historic-share-prices/.

37 Kraft Heinz lost: Kraft Heinz historical share price, February 16, 2018–February 18, 2020, retrieved from Kraft Heinz, http://ir.kraftheinzcompany.com/stock-information.

37 Its CEO has stepped down: Paul R. La Monica, "Kraft Heinz's Nightmare Is Far from Over," CNN Business, August 8, 2019, https://www.cnn.com/2019/08/08/investing/kraft-heinz-earnings/index.html.

37 it should be Unilever acquiring Kraft Heinz: Rob Cox, "Breakingviews—Cox: Unilever Can't Help but Mull a Kraft Pounce," March 7, 2019, https://www.reuters.com/article/us-kraft-heinz-unilever-breakingviews/breakingviews-cox-unilever-cant-help-but-mull-a-kraft-pounce-idUSKCN1QO22F.

37 "to be the best food company": "Growing a Better World at Kraft Heinz: 2017 Corporate Social Responsibility Report," 4, 5, 20, 22, 32, 54, 58, 66, 67, https://www.kraftheinzcompany.com/pdf/KHC_CSR_2017_Full.pdf.

CHAPTER 2: MAKING A KILLING

41 a single strand of hair: Curt Ellis, interview by Warren Valdmanis, June 2019. Unless otherwise noted, all remarks from Ellis are from this interview.

41 2.5 billion bushels: US Department of Agriculture, "Iowa Ag News—2018 Crop Production," February 8, 2019, https://www.nass.usda.gov/Statistics_by_State/Iowa/Publications/Crop_Report/2019/IA-Crop-Production-Annual-01-19.pdf.

42 Americans among the least healthy: Maggie Fox, "United States Comes in Last Again on Health, Compared to Other Countries," NBC News, November 16, 2016, https://www.nbcnews.com/health/health-care/united-states-comes-last-again-health-compared-other-countries-n684851.

42 one in five: Centers for Disease Control and Prevention, "Obesity," updated September 18, 2018, accessed January 13, 2020, https://www.cdc.gov/healthyschools/obesity/index.htm?CDC_AA_refVal=https%3A%2F%2Fwww.cdc.gov%2Fhealthyschools%2Fobesity%2Ffacts.htm.

42 Forty percent of American adults are obese: Centers for Disease Control and Prevention, "Adult Obesity Facts," updated August 13, 2018, accessed January 13, 2020, https://www.cdc.gov/obesity/data/adult.html.

42 Poor diet is now the leading cause: Ashkan Afshin, Patrick John Sur, Kairsten A. Fay, et al., "Health Effects of Dietary Risks in 195 Countries, 1990–2017: A Systematic Analysis for the Global Burden of Disease Study 2017," *Lancet* 393, no. 10184 (May 11, 2019): 1958–72, https://www.thelancet.com/article/S0140-6736(19)30041-8/fulltext; Kenneth D. Kochanek, Sherry L. Murphy, Jiaquan Xu, and Elizabeth Arias, "Deaths: Final Data for 2017," *National Vital Statistics Reports* 68, no. 9 (June 24, 2019), https://www.cdc.gov/nchs/data/nvsr/nvsr68/nvsr68_09-508.pdf.

43 That's 60 percent: American Heart Association, "Sugar Recommendation Healthy Kids and Teens Infographic," August 24, 2016, https://www.heart.org/en/healthy-living/healthy-eating/eat-smart/sugar/sugar-recommendation-healthy-kids-and-teens-infographic.

43 2 billion cans: "FAQs," Coca-Cola Journey, https://www.coca-colacompany.com/contact-us/faqs (site discontinued); also see https://web.archive.org/web/20160829091011/http://www.coca-colacompany.com:80/contact-us/faqs.

43 a million tacos in twenty-four hours: Sarah Yager, "Doritos Locos

Tacos: How Taco Bell and Frito-Lay Put Together One of the Most Successful Products in Fast-Food History," *The Atlantic*, July 2014, https://www.theatlantic.com/magazine/archive/2014/07/doritos-locos-tacos/372276/.

44 "a nineteenth-century catchall term": Howard Markel, *The Kelloggs: The Battling Brothers of Battle Creek* (New York: Pantheon, 2017), 35–36.

44 At the Battle Creek Sanitarium: Bhu Srinivasan, *Americana: A 400-Year History of American Capitalism* (New York: Penguin, 2017), 261.

45 Roughly 90 percent of Americans: "From the Breakfast Table to Snacking Staple: 43% of US Cereal Consumers Eat Cereal as a Snack at Home," Mintel, October 17, 2017, https://www.mintel.com/press-centre/food-and-drink/43-of-us-cereal-consumers-eat-cereal-as-a-snack-at-home.

46 published a paper: Michael C. Jensen and William H. Meckling, "Theory of the Firm: Managerial Behavior, Agency Costs, and Ownership Structure," *Journal of Financial Economics* 3, no. 4 (October 1976): 305–60, https://www.sciencedirect.com/science/article/pii/0304405X7690026X.

46 "Because it is logically impossible": Quoted in Justin Fox and Jay W. Lorsch, "What Good Are Shareholders?," *Harvard Business Review*, July–August 2012, 54.

47 With support from popular intellectuals: "In the 1950s," wrote the business scholar Jeffrey Pfeffer, "Milton Friedman was dismissed as a curiosity. By the 1980s, Friedman and several of his followers had won the Nobel Prize." Jeffrey Pfeffer, "Why Do Bad Management Theories Persist? A Comment on Ghoshal," *Academy of Management Learning & Education* 4, no. 1 (March 2005): 97.

47 "offered an easy-to-explain,": Lynn A. Stout, "New Thinking on 'Shareholder Primacy,'" *Accounting, Economics, and Law* 2, no. 2 (2012), https://scholarship.law.cornell.edu/cgi/viewcontent.cgi?article=2688&context=facpub, 3.

47 "No one was talking": Sheelah Kolhatkar, "The Economist Who Put Stock Buybacks in Washington's Crosshairs," *The New Yorker*, June 20, 2019, https://www.newyorker.com/business/currency/the-economist-who-put-stock-buybacks-in-washingtons-crosshairs.

47 Business research since the 1970s: Pfeffer, "Why Do Bad Management Theories Persist?"

47 "Shareholder primacy had become": Lynn A. Stout, "The Problem of Corporate Purpose," *Issues in Governance Studies* 48 (June 2012), https://www.brookings.edu/wp-content/uploads/2016/06/Stout_Corporate-Issues.pdf, 3.

47 the statements of the Business Roundtable: Claudine Gartenberg and George Serafeim, "181 Top CEOs Have Realized Companies Need a Purpose Beyond Profit," HBR.org, August 20, 2019, https://hbr.org/2019/08/181-top-ceos-have-realized-companies-need-a-purpose-beyond-profit.

48 "the legitimate concerns": Jeffrey N. Gordon, "The Rise of Independent Directors in the United States, 1950–2005: Of Shareholder Value and Stock Market Prices," *Stanford Law Review* 59 (2007): 1529, https://www.stanfordlawreview.org/print/article/the-rise-of-independent-directors-in-the-united-states-1950-2005-of-shareholder-value-and-stock-market-prices/.

48 "The principal objective": Ibid.

48 "The notion that the board": Duff McDonald, *The Golden Passport: Harvard Business School, the Limits of Capitalism, and the Moral Failure of the MBA Elite* (New York: Harper Business, 2017), 389. As we'll see, it is reversing itself again.

49 In one recent survey: David F. Larcker, Brian Tayan, Vinay Trivedi, and Owen Wurzbacher, "Stakeholders and Shareholders: Are Executives Really 'Penny Wise and Pound Foolish' About ESG?," *Stanford Closer Look Series,* July 2, 2019, https://www.gsb.stanford.edu/sites/gsb/files/publication-pdf/cgri-closer-look-78-esg-programs.pdf, 2.

49 The number of firms: *Short-Termism in Financial Markets: Hearing Before the Subcommittee on Economic Policy of the Committee on Banking, Housing, and Urban Affairs, United States Senate,* 111th Cong., 2nd sess., April 29, 2010, https://www.govinfo.gov/content/pkg/CHRG-111shrg61654/pdf/CHRG-111shrg61654.pdf, 27.

49 Two-thirds of CFOs: John C. Bogle, *The Battle for the Soul of Capitalism* (New Haven: Yale University Press, 2005), 99.

49 Four in five business leaders: Alfred Rappaport, "The Economics

of Short-Term Performance Obsession," *Financial Analysts Journal* 61, no. 3 (2005): 69.

49 Over half of managers surveyed: Miguel Padró, "Unrealized Potential: Misconceptions About Corporate Purpose and New Opportunities for Business Education," The Aspen Institute, May 28, 2014, https://assets.aspeninstitute.org/content/uploads/files/content/docs/pubs/Aspen%20BSP%20Unrealized%20Potential%20May2014v.2.pdf?_ga=2.40107429.603537148.1576800305-420537432.1576800305, 6.

47 "zombie idea": In that way, it is like the myth that people swallow eight spiders in their sleep each year, a claim that lacks any scientific backing yet is widely known. Ironically, the figure is so well known because it was published in a 1993 article as an example of the sort of false claim that could go viral. David Mikkelson, "Do People Swallow Eight Spiders Per Year?," Snopes, April 23, 2001, https://www.snopes.com/fact-check/swallow-spiders/.

49 "it is hard to find": Luigi Zingales, "Does Finance Benefit Society?," presidential address, American Finance Association Annual Meeting, Boston, MA, January 4, 2015, https://faculty.chicagobooth.edu/luigi.zingales/papers/research/finance.pdf, 13.

50 "Practical men": John Maynard Keynes, introduction to *The General Theory of Employment, Interest and Money* (London: Macmillan, 1936), 1, available at http://assets.press.princeton.edu/chapters/i9270.pdf.

50 "the road to heaven": Alan Atkisson, *Believing Cassandra: How to Be an Optimist in a Pessimist's World* (London: Routledge), 121.

51 "If empowering short-term investors": Leo E. Strine, Jr., "Who Bleeds When the Wolves Bite? A Flesh-and-Blood Perspective on Hedge Fund Activism and Our Strange Corporate Governance System," *Yale Law Journal* 126, no. 6 (2017): 1970.

52 almost all economic activity: Ralph Gomory and Richard Sylla, "The American Corporation," in "American Democracy & the Common Good," special issue, *Daedalus* 142, no. 2 (Spring 2013): 112.

52 More than 90 percent: Alan Greenspan and Adrian Wooldridge, *Capitalism in America: A History* (New York: Penguin, 2018), 32.

52 "every family in the country": "Thomas Jefferson to John Adams, 21 January 1812," Founders Online, National Archives, https://founders.archives.gov/documents/Jefferson/03-04-02-0334.

52 As late as 1850: Greenspan and Wooldridge, *Capitalism in America*, 427, 140.

52 There were no middle managers: Margaret M. Blair, "Locking in Capital: What Corporate Law Achieved for Business Organizers in the Nineteenth Century," *UCLA Law Review* 51, no. 2 (2003).

52 "the corporation undermined": Adam Smith, *An Inquiry into the Nature and Causes of the Wealth of Nations*, ed. Edwin Cannan (New York: Modern Library, 1994), 149, quoted in Thomas P. Byrne, "False Profits: Reviving the Corporation's Public Purpose," *UCLA Law Review* 57 (2010): 28–29.

52 As late as 1894: Sarah Pruitt, "When the Sears Catalog Sold Everything from Houses to Hubcaps," History, March 13, 2019, https://www.history.com/news/sears-catalog-houses-hubcaps.

53 had put together massive conglomerates: John Micklethwait and Adrian Wooldridge, *The Company: A Short History of a Revolutionary Idea* (New York: Modern Library, 2005), 66.

53 By 1860: Greenspan and Wooldridge, *Capitalism in America*, 433.

53 The number of white-collar workers: Ibid., 435.

53 the price of a bushel of wheat: Ibid., 120.

54 Before the Civil War: Srinivasan, *Americana*, 268.

54 A local, small, and simple industry: Ibid., 269.

54 Increasingly, however, they were owned: Stout, "The Problem of Corporate Purpose," 2.

54 The number of Americans: Micklethwait and Wooldridge, *The Company*, 117.

54 Four million workers staged: Greenspan and Wooldridge, *Capitalism in America*, 194.

54 By 1945, union membership: US Library of Congress, Congressional Research Service, *Union Membership Trends in the United States*, by Gerald Mayer, RL32553 (August 31, 2004), 12.

54 It also helped: Kim Phillips-Fein, *Invisible Hands: The Making of the Conservative Movement from the New Deal to Reagan* (New York: W. W. Norton, 2010), 88.

55 less than 7 percent: US Bureau of Labor Statistics, "Union Mem-

bers Summary," news release, January 18, 2019, https://www.bls.gov/news.release/union2.nr0.htm.

55 roughly two-thirds of public companies: Ronald J. Gilson, "Controlling Shareholders and Corporate Governance: Complicating the Comparative Taxonomy," *Harvard Law Review* 119, no. 6 (April 2006): 1646.

55 On the New York Stock Exchange: Ronald J. Gilson, "Controlling Shareholders and Corporate Governance: Complicating the Comparative Taxonomy," *Harvard Law Review* 119, no. 6 (April 2006): 1646.

55 The number of workers: Martin Gelter, "The Pension System and the Rise of Shareholder Primacy," Fordham Law Legal Studies Research Paper no. 2079607, July 22, 2012, 9.

56 After the war: Greenspan and Wooldridge, *Capitalism in America*, 273.

56 Real household income: Ibid., 274.

56 A 1961 survey revealed: "The Rise of Independent Directors," 1512.

56 "The position of shareholders": Quoted in ibid., 1514.

57 Through World War II: Greenspan and Wooldridge, *Capitalism in America*, 314.

57 The same was true of steel: Ibid., 315.

57 the economy stagnated: Lynn A. Stout, "The Toxic Side Effects of Shareholder Primacy," *University of Pennsylvania Law Review* 161, no. 7 (2013).

57 From 1966 through 1982: Ben Carlson, "Was the 1966–1982 Stock Market Really That Bad?," A Wealth of Common Sense, June 19, 2014, https://awealthofcommonsense.com/2014/06/1966-1982-stock-market-really-bad/.

57 Whereas in 1966: McDonald, *The Golden Passport*, 356.

57 nearly a quarter: Gordon, "The Rise of Independent Directors," 1521.

58 nearly three in five: Ronald J. Gilson and Jeffrey N. Gordon, "The Agency Costs of Agency Capitalism: Activist Investors and the Revaluation of Governance Rights," *Columbia Law Review* 113 (2013): 871.

58 By 2000, half: Micklethwait and Wooldridge, *The Company*, 129.

58 In 1981, 60 percent: Gelter, "The Pension System," 12.

58 Total assets in 401(k)s grew: Strine, "Who Bleeds When the Wolves Bite?," 1877.

58 In France, retirees get: Gelter, "The Pension System," 40.

59 In 1977, 20 percent: Gilson and Gordon, "The Agency Costs," 884.

59 By 2000, there were nine thousand: Greenspan and Wooldridge, *Capitalism in America*, 339.

59 Whereas institutions such as: Gilson and Gordon, "The Agency Costs," 874.

59 From 1973 to 1995: US Bureau of Economic Analysis, "Real Gross Domestic Product per Capita (A939RX0Q048SBEA)," Federal Reserve Bank of St. Louis, February 17, 2020, https://fred.stlouisfed.org/series/A939RX0Q048SBEA.

59 Productivity growth has slowed: Robert J. Gordon, *The Rise and Fall of American Growth: The U.S. Standard of Living Since the Civil War* (Princeton, NJ: Princeton University Press, 2016), 16, available at http://assets.press.princeton.edu/chapters/s10544.pdf.

60 become less dynamic: Jason Furman, "Prepared Testimony to the Hearing on 'Market Concentration,'" June 7, 2018, in a May 27, 2018, paper by Furman submitted to the Organisation for Economic Co-operation and Development Directorate for Financial and Enterprise Affairs Competition Committee, 6, https://one.oecd.org/document/DAF/COMP/WD(2018)67/en/pdf.

60 the number of public companies: Michael J. Mauboussin, Dan Callahan, and Darius Majd, "The Incredible Shrinking Universe of Stocks: The Causes and Consequences of Fewer U.S. Equities," Credit Suisse, March 22, 2017, https://research-doc.credit-suisse.com/docView?language=ENG&format=PDF&sourceid=em&document_id=1072753661&serialid=h%2B%2FwLdU%2FTIaitAx1rnamfYsPRAuTFRGdTSF4HZIvTkA%3D.

60 the company went public: Kellogg's, "FAQs," https://investor.kelloggs.com/FAQ.

60 Institutions now own: "Kellogg Company Common Stock (K) Institutional Holdings," Nasdaq, December 20, 2019, https://www.nasdaq.com/market-activity/stocks/k/institutional-holdings.

60 shareholders aren't even permitted: Kellogg's, "FAQs."

60 "I have been interested": Howard Markel, *The Kelloggs: The Battling Brothers of Battle Creek* (New York: Pantheon, 2017), 229.

60 Will pressed the doctor: Ibid., 137.

61 Kellogg's grew to be: "Kellogg Company Common Stock (K) Stock Quotes," Nasdaq, accessed December 20, 2019, https://www.nasdaq.com/symbol/k.

61 breakfast food each year: Kellogg Company, *2018 Annual Report, SEC Form 10-K and Supplemental Information*, 15, 16, 42, 104, http://www.annualreports.com/HostedData/AnnualReports/PDF/NYSE_K_2018.pdf.

61 "just collateral damage": Marion Nestle, *What to Eat* (New York: North Point Press, 2007), 13.

61 "Obesity is often": Joi Ito, "The World Is Complex. Measuring Charity Has to Be Too," *Wired*, July 30, 2019, https://www.wired.com/story/joi-ito-impact-investing/?mc_cid=cfa129c405&mc_eid=1e51a6314d.

62 unscrupulous financial advisers: For investment adviser rules, see Robert J. Jackson, Jr., "Statement on Final Rules Governing Investment Advice," Harvard Law School Forum on Corporate Governance and Financial Regulation, June 5, 2019, https://corpgov.law.harvard.edu/2019/06/05/statement-on-final-rules-governing-investment-advice/; and Vivek Bhattacharya, Gaston Illanes, and Manisha Padi, "Fiduciary Duty and the Market for Financial Advice," National Bureau of Economic Research Working Paper no. 25861, May 2019, https://www.nber.org/papers/w25861?utm_campaign=Hutchins%20Center&utm_source=hs_email&utm_medium=email&utm_content=72981934.

63 these companies control: Sophia Murphy, David Burch, and Jennifer Clapp, *Cereal Secrets: The World's Largest Grain Traders and Global Agriculture*, Oxfam International, August 2012, https://s3.amazonaws.com/oxfam-us/www/static/oa4/cereal-secrets.pdf.

63 "You can't have": Jon Olinto, interview by Warren Valdmanis, October 2019. Unless otherwise noted, all further remarks from Olinto are from this interview.

64 Kellogg's stock price: Kellogg's historical share price, January 2, 2014–December 31, 2019, retrieved from Kellogg's, https://

investor.kelloggs.com/Stock; S&P 500 historical daily price, January 2, 2014–December 31, 2019, retrieved from Cboe Global Markets, http://www.cboe.com/products/stock-index-options-spx-rut-msci-ftse/s-p-500-index-options/s-p-500-index/spx-historical-data.

64 What if, instead: See Nancy E. Roman, "The Power of Business to Change Food Culture for the Better," *Stanford Social Innovation Review* 17, no. 2 (Spring 2019), https://ssir.org/articles/entry/the_power_of_business_to_change_food_culture_for_the_better, and Harold Goldstein, "How Market Forces Could Improve How We Eat," *Stanford Social Innovation Review* 17, no. 3 (Summer 2019), https://ssir.org/articles/entry/how_market_forces_could_improve_how_we_eat.

65 "This is rat eat rat": Eric Schlosser, *Fast Food Nation: The Dark Side of the All-American Meal* (New York: Houghton Mifflin, 2001), 37.

65 "that ethics, like the caboose": McDonald, *The Golden Passport,* 435, citing Laurence Shames, *The Big Time: The Harvard Business School's Most Successful Class and How It Shaped America* (New York: Harper & Row, 1986), 175.

65 "To educate a man": James H. Billington, *Respectfully Quoted: A Dictionary of Quotations* (Mineola, NY: Dover Publications, 2010), 99.

CHAPTER 3: GOOD NEWS AND BAD NEWS

67 "We now open": Gannett annual meeting, May 13, 2019, McLean, Virginia, attended by Michael O'Leary. Unless otherwise noted, all further descriptions of the proceedings and quotes from attendees have been supplied by the author.

68 "one of the most": Margaret Sullivan, "Is This Strip-Mining or Journalism? 'Sobs, Gasps, Expletives' over Latest *Denver Post* Layoffs," *Washington Post,* March 15, 2018, https://www.washingtonpost.com/lifestyle/style/is-this-strip-mining-or-journalism-sobs-gasps-expletives-over-latest-denver-post-layoffs/2018/03/15/d05abc5a-287e-11e8-874b-d517e912f125_story.html.

68 "Your newspaper-killing business model": Sherrod Brown (@SenSherrodBrown), "Local newspapers form the foundation of our country," open letter to Heath Bradford Freeman and

Randall Duncan Smith, Twitter, April 25, 2019, 3:12 p.m., https://twitter.com/SenSherrodBrown/status/1121537591231373317/photo/2.

69 that's 15 million more people: 122 million people voted in the 2018 elections (Jens Manuel Krogstad, Luis Noe-Bustamante, and Antonio Flores, "Historic Highs in 2018 Voter Turnout Extended Across Racial and Ethnic Groups," Pew Research Center, May 1, 2019), whereas 54 percent of adults owned stock directly or indirectly in 2009–2017 (Jeffrey M. Jones, "U.S. Stock Ownership Down Among All but Older, Higher-Income," Gallup, May 24, 2017). There were 327 million people in the United States when the 2018 elections took place, of whom 77.6 percent were over the age of eighteen (US Census Bureau, "U.S. and World Population Clock," https://www.census.gov/popclock/), which implies that 137 million people own stock directly or indirectly.

69 the wealthiest 1 percent hold: Danielle Kurtzleben, "While Trump Touts Stock Market, Many Americans Are Left Out of the Conversation," NPR, March 1, 2017, https://www.npr.org/2017/03/01/517975766/while-trump-touts-stock-market-many-americans-left-out-of-the-conversation; Patricia Cohen, "We All Have a Stake in the Stock Market, Right? Guess Again," *New York Times*, February 8, 2018, https://www.nytimes.com/2018/02/08/business/economy/stocks-economy.html.

69 a retirement account worth $65,000: William W. Bratton and Michael L. Wachter, "Shareholders and Social Welfare," *Seattle University Law Review* 36, no. 2 (Winter 2013): 510.

70 "to rethink its business strategy": Denver Post Editorial Board, "Editorial: As Vultures Circle, The Denver Post Must Be Saved," *Denver Post*, April 6, 2018, https://www.denverpost.com/2018/04/06/as-vultures-circle-the-denver-post-must-be-saved/.

70 "If we don't": Ricardo Baca, "When a Hedge Fund Tries to Kill the Newspapers It Owns, Journalists Must Fight Back," *Denver Post*, April 6, 2018, https://www.denverpost.com/2018/04/06/when-a-hedge-fund-tries-to-kill-the-newspapers-it-owns-journalists-must-fight-back/.

70 John Louis had met: Gannett/*USA Today* Network, Schedule 14A Proxy Statement, US Securities and Exchange Commission,

March 26, 2019, https://www.sec.gov/Archives/edgar/data/1635718/000119312519085918/d601169ddefc14a.htm.

71 7.5 percent of Gannett's stock: Cara Lombardo, "Hedge-Fund-Backed Media Group Makes Bid for Gannett," *Wall Street Journal*, January 14, 2019, https://www.wsj.com/articles/hedge-fund-backed-media-group-prepares-bid-for-gannett-11547427720.

71 "Within the industry": "Gannett Board of Directors Unanimously Rejects Unsolicited Proposal from MNG Enterprises, Inc.," Gannett, February 4, 2019, http://ir.newmediainv.com/Cache/1001258709.PDF?O=PDF&T=&Y=&D=&-FID=1001258709&iid=4426551.

71 "a home for the Company's": Rick Edmonds, "Gannett Rejects Acquisition Bid, Says Digital First Would Be Unfit to Run Its Properties," Poynter, February 4, 2019, https://www.poynter.org/business-work/2019/gannett-rejects-acquisition-bid-says-digital-first-would-be-unfit-to-run-its-properties/.

71 Gannett's stock price: MNG Enterprises, "An Opportunity for Immediate and Compelling Value for Gannett Shareholders," April 9, 2019, 25, https://www.sec.gov/Archives/edgar/data/1635718/000092189519001065/investorpresentation.pdf.

72 "No margin, no mission": John Paul Newport, "Health Care; Mission + Margin: The Nun as C.E.O.," *New York Times*, June 9, 1991, https://www.nytimes.com/1991/06/09/magazine/health-care-mission-margin-the-nun-as-ceo.html.

72 Daily subscriptions nationwide: Michael Barthel, "Newspapers Fact Sheet," Pew Research Center, July 9, 2019, https://www.journalism.org/fact-sheet/newspapers/.

72 Annual revenue: Marc Tracy, "A Paradox at the Heart of the Newspaper Crisis," *New York Times*, August 1, 2019, https://www.nytimes.com/2019/08/01/business/media/news-deserts-media-newspapers.html.

73 the *Tribune* would become a nonprofit: Katherine Rosman, "Can Paul Huntsman Save the Salt Lake Tribune?," *New York Times*, May 17, 2019, https://www.nytimes.com/2019/05/17/business/media/paul-huntsman-salt-lake-tribune.html.

73 the *New York Times* earned: Jonathan Shieber, "Profits at The

New York Times Show Media Dinosaurs Are Ruling the Internet," TechCrunch, February 6, 2019, https://techcrunch.com/2019/02/06/profits-at-the-new-york-times-show-media-dinosaurs-are-ruling-the-internet/.

73 *The Guardian* earned a profit: Joshua Benton, "Want to See What One Digital Future for Newspapers Looks Like? Look at *The Guardian,* Which Isn't Losing Money Anymore," Nieman Lab, May 1, 2019, https://www.niemanlab.org/2019/05/want-to-see-what-one-digital-future-for-newspapers-looks-like-look-at-the-guardian-which-isnt-losing-money-anymore/.

73 It now makes: Ibid.

73 Over a single year: Ibid.

74 Vox, Vice, Axios, The Information, and others: Though even these can struggle. The old law remains: *No margin, no mission.* "Disney wrote down $353m from its stake in Vice Media, another warning sign for the once-mighty digital media company." Anna Nicolaou, "Disney Writes Down $353m from Vice Stake," *Financial Times,* May 8, 2019, https://www.ft.com/content/9f4988de-71df-11e9-bf5c-6eeb837566c5.

75 "I didn't want it to shut down": Rich Klein, interview by Michael O'Leary, May 2, 2019. Unless otherwise noted, all further remarks from Klein are from this interview.

75 "The main thing": Clay Lambert, interview by Michael O'Leary, May 1, 2019. Unless otherwise noted, all further remarks from Lambert are from this interview.

76 That was still true: 1,300 of the 1,785 local papers in 1953. David Collis, Peter Olson, and Mary Furey, "The Newspaper Industry in Crisis," Harvard Business School, 9-709-463 (January 12, 2010): 5–6.

76 only 15 percent of daily newspapers: Dirks, Van Essen, Murray & April, "History of Ownership Consolidation," press release, March 31, 2017, http://dirksvanessen.com/articles/view/223/history-of-ownership-consolidation-/.

76 Gannett itself bought: Ibid.

76 59 percent of Americans: Richard Reeves, "Starbucks v Dunkin': How Capitalism Gives Us the Illusion of Choice," *The Guardian,*

July 29, 2019, https://www.theguardian.com/commentisfree/2019/jul/29/starbucks-v-dunkin-how-capitalism-gives-us-the-illusion-of-choice.

77 75 percent of the general public: John C. Bogle, *The Battle for the Soul of Capitalism* (New Haven: Yale University Press, 2005), 22.

77 "everybody in the local community": Wendell Berry, "The Idea of a Local Economy," *Orion Magazine*, January 1, 2001, https://orionmagazine.org/article/the-idea-of-a-local-economy/.

78 three days after: All this, unless otherwise noted, is from Gannett, "2019 Proxy Statement," March 12, 2019, https://www.sec.gov/Archives/edgar/data/1635718/000119312519085918/d601169ddefc14a.htm.

79 "expressing gratitude for arranging the meeting": Gannett, "2019 Proxy Statement," 12. The quote is from the proxy, not the email.

80 All public companies are run: This is no accident. The governance of the Dutch East India Company was based on the Dutch Republic, and the governance of the English East India Company was based on the House of Commons with a constitution, Parliament, and prime minister mirrored in the shareholders, charter, board, and chairman of the board. See David A. Ciepley, "Beyond Public and Private: Toward a Political Theory of the Corporation," *American Political Science Review* 107, no. 1 (February 2013): 141–42.

80 63 percent by 2016: Lucian A. Bebchuk, Alma Cohen, and Scott Hirst, "The Agency Problems of Institutional Investors," *Journal of Economic Perspectives* 31, no. 3 (Summer 2017): 91, https://pubs.aeaweb.org/doi/pdfplus/10.1257/jep.31.3.89.

80 institutions now control: Ronald J. Gilson and Jeffrey N. Gordon, "The Agency Costs of Agency Capitalism: Activist Investors and the Revaluation of Governance Rights," *Columbia Law Review* 113 (2013): 874.

80 93 percent are cast by institutions: Einer R. Elhauge, "The Causal Mechanisms of Horizontal Shareholding," August 2, 2019, available at SSRN: https://ssrn.com/abstract=3370675 or http://dx.doi.org/10.2139/ssrn.3370675.

80 Who are these nameless: "Associate Members," Council of Institutional Investors, October 30, 2019, https://www.cii.org/associate_members.

80 entrusted to them by investors: Suzy Waite, Annie Massa, and Christopher Cannon, "Asset Managers with $74 Trillion on Brink of Historic Shakeout," Bloomberg, August 8, 2019, https://www.bloomberg.com/graphics/2019-asset-management-in-decline/; State Street, "State Street Reports Fourth-Quarter 2018 EPS of $1.04," press release, January 18, 2019, https://newsroom.statestreet.com/press-release/corporate/state-street-reports-fourth-quarter-2018-eps-104-eps-168-excluding-notable-i.

80 Together, they are the largest: Jan Fichtner and Eelke Heemskerk, "The New Permanent Universal Owners: Index Funds, (Im)patient Capital, and the Claim of Long-Termism," SSRN, November 13, 2018, 6.

81 "It is not an exaggeration": John C. Coates, "The Future of Corporate Governance, Part I: The Problem of Twelve," Harvard Public Law Working Paper No. 19-07, September 20, 2018, https://corpgov.law.harvard.edu/wp-content/uploads/2019/11/John-Coates.pdf, 14.

81 These institutional investors: Alfred Rappaport, "The Economics of Short-Term Performance Obsession," *Financial Analysts Journal* 61, no. 3 (2005): 66.

81 "foster informed and engaged": Knight Foundation homepage, https://knightfoundation.org/about/.

82 At BlackRock, for example: Fichtner and Heemskerk, "The New Permanent Universal Owners," 5.

82 Two firms: Leo E. Strine Jr., "Who Bleeds When the Wolves Bite? A Flesh-and-Blood Perspective on Hedge Fund Activism and Our Strange Corporate Governance System," *The Yale Law Journal* 126, no. 6 (2017), https://scholarship.law.upenn.edu/cgi/viewcontent.cgi?article=2729&context=faculty_scholarship, 1904.

82 An ISS recommendation: Paul H. Edelman, Randall S. Thomas, and Robert B. Thompson, "Shareholder Voting in an Age of Intermediary Capitalism," *Southern California Law Review* 87, no. 6 (September 2014): 1361.

82 "This is, of course": Leo E. Strine Jr., "Can We Do Better by Ordinary Investors?: A Pragmatic Reaction to the Dueling Ideological

Mythologists of Corporate Law," *Columbia Law Review* 114, no. 449 (2014): 478, https://millstein.law.columbia.edu/sites/default/files/content/docs/Strine%202014%20Columbia%20LR.%20Dueling%20Ideologues.pdf.

82 State Street offers: Vanguard votes the shares of its Social Index Fund the exact same way as it does every other one of its funds, whether or not the proposal in question deals with social issues.

83 Milgram-Nixon Syndrome: This idea was originally phrased as the "Eichmann-Nixon Syndrome" by Bruce Mathers, based on Eichmann's defense at the Nuremberg trials. We've altered the phrase because we think Milgram's experiments are more appropriate to the subject matter. Bruce Mathers, "The Eichmann-Nixon Syndrome Bedevils Large Companies," *Financial Times*, November 12, 2018, https://www.ft.com/content/26f6e572-e678-11e8-8a85-04b8afea6ea3.

83 series of experiments: Cari Romm, "Rethinking One of Psychology's Most Infamous Experiments," *The Atlantic*, January 28, 2015, https://www.theatlantic.com/health/archive/2015/01/rethinking-one-of-psychologys-most-infamous-experiments/384913/.

84 this state of mind: American Psychological Association, *Dictionary of Psychology*, definition of "agentic state," https://dictionary.apa.org/agentic-state.

85 Typical shareholders are fifty-one years old: Bratton and Wachter, "Shareholders and Social Welfare," 510.

86 Almost all of their financial savings: Ibid.

86 They can't even access: "Early Retirement Distribution Options," 401khelpcenter.com, http://www.401khelpcenter.com/401k_education/Early_Dist_Options.html.

86 Someone that age: "Actuarial Life Table," Social Security Administration, March 15, 2006, https://www.ssa.gov/oact/STATS/table4c6.html.

86 For example, someone who invests: "Vanguard Target Retirement Funds," Vanguard Group, https://investor.vanguard.com/mutual-funds/target-retirement/.

86 Index funds like these: Coates, "The Future of Corporate Governance," 13.

86 90 percent are invested: Bratton and Wachter, "Shareholders and Social Welfare," 510; Strine, "Who Bleeds When the Wolves Bite?," 1878.

86 $65,000 worth of financial assets: Bratton and Wachter, "Shareholders and Social Welfare," 510.

86 For most shareholders: Strine, "Who Bleeds When the Wolves Bite?," 1877.

87 "the failure of our overall": Ibid., 1874.

87 these are the shareholders: The economists Oliver Hart and Luigi Zingales have made a similar argument for corporations serving shareholder *welfare* rather than just shareholder value. This includes accounting for shareholders' ethical and social concerns. Oliver Hart and Luigi Zingales, "Companies Should Maximize Shareholder Welfare Not Market Value," *Journal of Law, Finance and Accounting* 2 (2017): 247–74. See also Oliver Hart and Luigi Zingales, "Serving Shareholders Doesn't Mean Putting Profit Above All Else," *Harvard Business Review*, October 12, 2016, https://hbr.org/2017/10/serving-shareholders-doesnt-mean-putting-profit-above-all-else.

88 "Fishing with dynamite": Lynn A. Stout, *The Shareholder Value Myth: How Putting Shareholders First Harms Investors, Corporations, and the Public* (San Francisco: Berrett-Koehler, 2012), 51.

89 Digital First didn't win: Cara Lombardo, "Gannett Poised to Hold Off Digital First Takeover Bid," MarketWatch, May 15, 2019, https://www.marketwatch.com/story/gannett-poised-to-hold-off-digital-first-takeover-bid-2019-05-15.

90 "It's not something": John Lauve, interview by Michael O'Leary, May 16, 2019. Unless otherwise noted, all further remarks from Lauve are from this interview.

CHAPTER 4: FIGHTING FOR CORPORATE SOCIAL RESPONSIBILITY

91 "We were providing": Interview with former Lehman Brothers executive by Warren Valdmanis, November 2018. Unless otherwise noted, all further remarks from the executive are from this interview.

92 "When the music stops": Michiyo Nakamoto, "Citigroup Chief

Stays Bullish on Buyouts," *Financial Times,* July 9, 2007, https://www.ft.com/content/80e2987a-2e50-11dc-821c-0000779fd2ac.

92 "Great rallies were staged": Lawrence G. McDonald, *A Colossal Failure of Common Sense: The Inside Story of the Collapse of Lehman Brothers* (New York: Crown Business, 2009), chap. 9.

93 Eighty-six percent: Governance and Accountability Institute, "Flash Report: 86% of S&P 500 Index® Companies Publish Sustainability/Responsibility Reports in 2018," May 16, 2019, https://www.ga-institute.com/press-releases/article/flash-report-86-of-sp-500-indexR-companies-publish-sustainability-responsibility-reports-in-20.html.

93 the average company spends: CECP in association with The Conference Board, "Giving in Numbers: 2018 Edition," 7, https://cecp.co/wp-content/uploads/2018/12/GIN2018_web.pdf.

93 combined annual revenues: "Global 500," *Fortune,* as of February 20, 2020, fortune.com, https://fortune.com/global500/.

93 US gross domestic product: US Bureau of Economic Analysis, Gross Domestic Product, Federal Reserve Bank of St. Louis, February 18, 2020, https://fred.stlouisfed.org/series/GDP.

93 invest $100 million: Chevron, "Chevron Technology Ventures Launches Future Energy Fund," press release, June 20, 2018, https://www.chevron.com/stories/chevron-technology-ventures-launches-future-energy-fund.

93 capital and exploratory investments: Chevron, *2018 Annual Report,* 39, https://www.chevron.com/-/media/chevron/annual-report/2018/documents/2018-Annual-Report.pdf.

94 an award in 2008: Saba Salman, "The Pin-Striped Philanthropists," *The Guardian,* November 4, 2008, https://www.theguardian.com/society/2008/nov/05/philanthropy-city-lehman-brothers-charity.

94 "a great vampire squid": Matt Taibbi, "The Great American Bubble Machine," *Rolling Stone,* April 5, 2010, https://www.rollingstone.com/politics/politics-news/the-great-american-bubble-machine-195229/.

95 In partnership with the World Bank: Goldman Sachs, "IFC, Goldman Sachs Initiative Invests $1 Billion in Women Entrepreneurs in Emerging Markets," press release, May 18, 2018, https://www

.goldmansachs.com/media-relations/press-releases/current/ifc-gs-investment-18-may-2018.html.

95 catering business in Lagos: "Meet the Women," 10,000 Women, Goldman Sachs, https://www.goldmansachs.com/citizenship/10000women/meet-the-women-profiles/ayodeji-profile.html.

95 "As a financial services company": Goldman Sachs (@GoldmanSachs), Twitter, March 8, 2019, 2:30 p.m., https://twitter.com/goldmansachs/status/1104102071509401600, citing Goldman Sachs, "When Women Lead," https://www.goldmansachs.com/our-firm/people-and-culture/when-women-lead/index.html?cid=scl%2Dnp%2Dtwitter%2Devergreen%2D-post%2D20171&sf208995379=1.

95 eleven-person board of directors: Bradley Keoun, "The Goldman Sachs Board Remains Old Boys' Club Even as Rivals Promote Women," *The Street*, May 1, 2018, https://www.thestreet.com/investing/as-companies-add-more-women-to-boards-goldman-sachs-keeps-a-pair-14574319.

95 358th out of the Fortune 500: Ibid.

95 thirty-person management committee: Ibid.

95 earn 55 percent less: "Goldman Sachs Reports Gender Pay Gap of 55.5 Percent," Reuters, March 16, 2018, https://www.reuters.com/article/us-goldman-sachs-pay-gender/goldman-sachs-reports-gender-pay-gap-of-55-5-percent-idUSKCN1GS1T3.

95 more men than women in senior positions: Ibid.

95 in 2018 alone: Goldman Sachs, "Goldman Sachs Reports Earnings per Common Share of $25.27 for 2018: Full Year and Fourth Quarter 2018 Earnings Results," press release, January 16, 2019, https://www.goldmansachs.com/media-relations/press-releases/current/pdfs/2018-q4-results.pdf.

96 salaries and benefits: The Goldman Sachs Group Inc., *2018 Annual Report*, 54, https://www.goldmansachs.com/investor-relations/financials/current/annual-reports/2018-annual-report/annual-report-2018.pdf.

96 Goldman announced: Julia Horowitz, "Goldman Sachs Says It Must Hire More Women and Minorities," CNN Business, March 18, 2019, https://www.cnn.com/2019/03/18/business/goldman-sachs-diversity-hiring-plan/index.html.

96 "influential, positive force": "Purpose & Progress: 2017 Environmental, Social and Governance Report," Goldman Sachs, 3, https://www.goldmansachs.com/citizenship/sustainability-reporting/esg-content/esg-report-2017.pdf.

96 the terms "gender," "women": The Goldman Sachs Group, Inc., *2017 Annual Report,* 190, https://www.sec.gov/Archives/edgar/data/886982/000119312518056383/d480167d10k.htm.

96 "with purposes, duties": William T. Allen, "Our Schizophrenic Conception of the Corporation," *Cardozo Law Review* 14, no. 261 (1992): 270.

97 half of the companies: "ESG Revolution Rising: From Low Chatter to Loud Roar," Goldman Sachs, April 23, 2018, https://www.goldmansachs.com/insights/pages/esg-revolution-rising.html.

97 96 percent say: David F. Larcker et al., "Stakeholders and Shareholders: Are Executives Really 'Penny Wise and Pound Foolish' about ESG?," *Stanford Closer Look Series* (July 2019): 2.

97 "The movement for corporate": Daniel J. Morrissey, "The Riddle of Shareholder Rights and Corporate Social Responsibility," *Brooklyn Law Review* 80, no. 2 (2015): 384.

97 the typical episode: Alex McLevy, "Happy 10th Anniversary to *Undercover Boss,* the Most Reprehensible Propaganda on TV," *A.V. Club,* February 5, 2020, https://tv.avclub.com/happy-10th-anniversary-to-undercover-boss-the-most-rep-1841278475.

98 "Every time you buy": "Hope Detector," Hyundai NFL Super Bowl LII commercial, aired February 4, 2018, https://www.youtube.com/watch?v=yihgufUn86g.

98 have both been featured: Kellogg/Cision PR Newswire, "Kellogg Company Named One of the World's Most Ethical Companies by Ethisphere for the 11th Time," press release, February 26, 2019, https://www.prnewswire.com/news-releases/kellogg-company-named-one-of-the-2019-worlds-most-ethical-companies-by-ethisphere-for-the-11th-time-300802327.html; Nina Lentini, "McDonald's Is Among US Firms Named 'World's Most Ethical' by Magazine," *MediaPost,* May 11, 2007, https://www.mediapost.com/publications/article/60176/mcdonalds-is-among-us-firms-named-worlds-most.html.

99 first Fortune 500 company: The Coca-Cola Company, *2018 Busi-*

ness & Sustainability Report, April 24, 2019, https://www.coca-colacompany.com/content/dam/journey/us/en/private/fileassets/pdf/2019/Coca-Cola-Business-and-Sustainability-Report.pdf.

99 The average American: Eliza Barclay, Julia Belluz, and Javier Zarracina, "It's Easy to Become Obese in America. These 7 Charts Explain Why," Vox, August 9, 2018, https://www.vox.com/platform/amp/2016/8/31/12368246/obesity-america-2018-charts.

99 "Black and Hispanic consumers": Jennifer L. Harris and Willie Frazier III, *Rudd Report*, UConn Rudd Center for Food Policy & Obesity, January 2019, 4, http://uconnruddcenter.org/files/Pdfs/TargetedMarketingReport2019.pdf.

99 Coca-Cola executives: Anahad O'Connor, "Coca-Cola Funds Scientists Who Shift Blame for Obesity Away from Bad Diets," *New York Times*, August 9, 2015, https://well.blogs.nytimes.com/2015/08/09/coca-cola-funds-scientists-who-shift-blame-for-obesity-away-from-bad-diets/.

99 also grew its sales of water: Luis Sanchez, "How Coca-Cola Is Thriving Despite Declining Soda Consumption," The Motley Fool, February 12, 2019, updated April 15, 2019, accessed January 15, 2020, https://www.fool.com/investing/2019/02/12/how-coca-cola-is-thriving-despite-declining-soda-c.aspx.

100 he offered a promotion: James O'Toole, *The Enlightened Capitalists: Cautionary Tales of Business Pioneers Who Tried to Do Well by Doing Good* (New York: Harper Business, 2019), 54.

100 With $6 trillion of assets: Dawn Lim, "BlackRock's Profit Declines, but Assets Under Management Again Top $6 Trillion," *Wall Street Journal*, April 16, 2019, https://www.wsj.com/articles/blackrocks-profit-declines-but-assets-under-management-again-top-6-trillion-11555411804.

101 a longer letter: Larry Fink, "A Sense of Purpose," letter to CEOs, January 16, 2018, https://www.blackrock.com/corporate/investor-relations/2018-larry-fink-ceo-letter.

101 CVS changed its approach: Larry J. Merlo, president and CEO, CVS, video message on tobacco sales, February 5, 2014, https://cvshealth.com/thought-leadership/message-from-larry-merlo-president-and-ceo.

102 Dollar stores had recently introduced: "Dollar-Store Chains Find Smokers Are Some of Their Best Shoppers," Newsmax, November 5, 2014, https://www.newsmax.com/Finance/dollar-store-retail-tobacco-cigarette/2014/11/05/id/605459/; "Dollar General Adds Tobacco," Midwest Independent Retailers Association, December 12, 2012, https://miramw.org/business-news/dollar-general-adds-tobacco/.

102 When CVS made the change: Patrick Grubbs, "CVS Quit Cigarettes So Americans Would Too—and It Worked," TriplePundit, May 28, 2019, https://www.triplepundit.com/story/2019/cvs-quit-cigarettes-so-americans-would-too-and-it-worked/83651.

102 Total cigarette sales declined: Ronnie Cohen, "When CVS Stopped Selling Cigarettes, Some Customers Quit Smoking," Reuters, March 20, 2017, https://www.reuters.com/article/us-health-pharmacies-cigarettes/when-cvs-stopped-selling-cigarettes-some-customers-quit-smoking-idUSKBN16R2HY.

102 "As a health insurer": "AXA Signs the Tobacco Free Finance Pledge," AXA, September 26, 2018, https://www.axa.com/en/newsroom/news/axa-signs-the-tobacco-free-finance-pledge.

102 "Insurers should always be": C. B. Bhattacharya, *Small Actions, Big Difference: Leveraging Corporate Sustainability to Drive Business and Societal Value* (New York: Routledge, 2020), chap. 3; AXA, "A First Step to Fight Tobacco," May 31, 2016, https://www.axa.com/en/newsroom/news/a-first-step-to-fight-tobacco.

102 "It was nonsense": Céline Soubranne, interview by Michael O'Leary, August 27, 2019. Unless otherwise noted, all further remarks from Soubranne are from this interview.

103 Wells Fargo announced: Wells Fargo, "Since 2015, Wells Fargo Has Donated $100 Million to Nonprofits Serving People with Disabilities," CSRwire, July 17, 2019, https://www.csrwire.com/press_releases/42219-Since-2015-Wells-Fargo-Has-Donated-100-Million-to-Nonprofits-Serving-People-with-Disabilities.

104 In 2018, only one company: Larcker et al., "Stakeholders and Shareholders," 5.

104 But between a company's: Florian Berg, Julian F. Koelbel, and Roberto Rigobon, "Aggregate Confusion: The Divergence of ESG

Ratings," MIT Sloan School of Management Working Paper No. 5822-19, August 15, 2019, 2.

104 "Investors should not treat": James Mackintosh, "Is Tesla or Exxon More Sustainable? It Depends Whom You Ask," *Wall Street Journal*, September 17, 2018, https://www.wsj.com/articles/is-tesla-or-exxon-more-sustainable-it-depends-whom-you-ask-1537199931.

104 four in five CEOs: SSRS, "A View from the Top: U.S. Fortune 1000 CEOs and C-Suite Executives on Social Purpose and Its Impact on Business," Covestro i³ Index, based on phone interviews with 100 executives conducted October 26, 2017–January 16, 2018, https://www.covestro.us/en/csr-and-sustainability/i3/covestro-i3-index.

105 the reasons ratings differ: Tracy Mayor, "Why ESG Ratings Vary So Widely. And What You Can Do About It," August 26, 2019, MIT Sloan School of Management, https://mitsloan.mit.edu/ideas-made-to-matter/why-esg-ratings-vary-so-widely-and-what-you-can-do-about-it.

105 SASB wants companies: "SASB Materiality Map," Sustainability Accounting Standards Board, February 12, 2015, https://materiality.sasb.org/.

106 We know what: JUST Capital and its survey of more than 96,000 Americans is a useful resource for this. JUST Capital, "A Roadmap for Stakeholder Capitalism: JUST Capital's 2019 Survey Results," October 2019, https://justcapital.com/2019-survey-report-download/.

107 "I think there's increasing": "For Mary Schapiro, Sustainability Disclosures Are Gaining Ground," *Wall Street Journal*, December 20, 2016, https://deloitte.wsj.com/riskandcompliance/2016/12/20/for-mary-schapiro-sustainability-disclosures-are-gaining-ground/.

107 put Philips in the ninety-eighth percentile: Koninklijke Philips Electronics Na CSR/ESG Ranking, CSRHub, https://www.csrhub.com/CSR_and_sustainability_information/Koninklijke-Philips-Electronics-Na.

107 *Fortune* magazine ranked Philips: Philips Group, "Philips Ranks #1

on Fortune's Change the World Sustainability All Stars List," August 21, 2019, https://www.philips.com/a-w/about/news/archive/standard/news/articles/2019/20190821-philips-ranks-1-on-fortunes-change-the-world-sustainability-all-stars-list.html.

108 its annual report to shareholders: Philips Group, *Annual Report 2018*, February 26, 2019, https://www.philips.com/c-dam/corporate/about-philips/sustainability/downloads/other/philips-full-annual-report-2018.pdf.

108 And it has the results: Edward Walsh, interview by Michael O'Leary, December 2, 2019. Unless otherwise noted, all further remarks from Walsh are from this interview.

108 $30 billion investment firm: Michael Ellis (COO/CCO, Inherent Group), email message to Michael O'Leary, February 11, 2020.

108 "I'd become": Tony Davis, interview by Michael O'Leary, December 6, 2019. Unless otherwise noted, all further remarks from Davis are from this interview.

109 launched a new investment firm: Inherent Group, "Strategy," https://www.inherentgroup.com/strategy/.

109 Imagine you're running a restaurant: Peter Thiel with Blake Masters, *Zero to One: Notes on Startups, or How to Build the Future* (New York: Crown Business, 2014), 31.

109 Google's market share: Jeff Desjardins, "How Google Retains More than 90% of Market Share," *Business Insider*, April 23, 2018, https://www.businessinsider.com/how-google-retains-more-than-90-of-market-share-2018-4.

109 "characteristic of a kind": Thiel, *Zero to One*, 31.

110 Meanwhile, Mary Barra: Michael Wayland, "UAW Members Approve Labor Deal to End Strike with GM; Union Selects Ford Next," CNBC, October 25, 2019, https://www.cnbc.com/2019/10/25/uaw-members-approve-new-labor-contract-with-gm-ending-40-day-strike.html.

110 Though GM has: Revenue stats as of 2018; market cap stats as of December 3, 2019. "Salesforce.com, Inc. (CRM) Income Statement," Yahoo! Finance, https://finance.yahoo.com/quote/CRM/financials?p=CRM; "General Motors Company (GM) Income Statement," Yahoo! Finance, https://finance.yahoo.com/quote/GM/financials?p=GM.

110 The researchers found: Deutsche Asset & Wealth Management (UK) Limited and the University of Hamburg (Germany), with contributors Gunnar Friede, Michael Lewis, Alexander Bassen, and Timo Busch and a foreword by Fiona Reynolds, "ESG & Corporate Financial Performance: Mapping the Global Landscape," December 3, 4, 2015, https://institutional.dws.com/content/_media/K15090_Academic_Insights_UK_EMEA_RZ_Online_151201_Final_(2).pdf.

111 "evidence that *High Sustainability*": Robert G. Eccles, Ioannis Ioannou, and George Serafeim, "The Impact of Corporate Sustainability on Organizational Processes and Performance," *Management Science* 60, no. 11 (February 2014): 1, https://www.hbs.edu/faculty/Publication%20Files/SSRN-id1964011_6791edac-7daa-4603-a220-4a0c6c7a3f7a.pdf.

112 The salad restaurant MIXT: "Our Story," MIXT, last modified September 20, 2018, accessed December 20, 2019, https://www.mixt.com/our-story/.

112 "When the patient has cancer": Paul Adler, "Stewardship Versus the Market: A Skeptical Perspective," in *Corporate Stewardship: Achieving Sustainable Effectiveness*, ed. Susan Albers Mohrman, James O'Toole, and Edward E. Lawler III (New York: Routledge, 2017), 255.

113 The Council represents: Council of Institutional Investors, "About CII," https://www.cii.org/about.

114 "CEOs are themselves": Wall Street Journal Editorial Board, "The 'Stakeholder' CEOs," *Wall Street Journal*, August 19, 2019, https://www.wsj.com/articles/the-stakeholder-ceos-11566248641.

114 "Accountability to everyone": Council of Institutional Investors, "Council of Institutional Investors Responds to Business Roundtable Statement on Corporate Purpose," August 19, 2019, https://www.cii.org/aug19_brt_response.

CHAPTER 5: HEAR NO EVIL, SEE NO EVIL

117 "Disclose, divest": Alexandra A. Chaidez and Luke W. Vrotsos, "A New Day for Divestment," *Harvard Crimson*, May 28, 2019, https://www.thecrimson.com/article/2019/5/28/a-new-day-for-divestment/.

118 "research, education": Alexandra A. Chaidez and Aidan F. Ryan, "Divest Protestors Pay Bacow a Morning Visit at Elmwood," *Harvard Crimson*, September 6, 2019, https://www.thecrimson.com/article/2019/9/6/divest-pancakes-at-elmwood/.

118 "The University maintains": "Shareholder Responsibility Committees," Harvard University, https://www.harvard.edu/shareholder-responsibility-committees.

118 "We condemn investment": Fossil Fuel Divest Harvard, http://divestharvard.com/.

118 "The faculty and students": Ned Hall, interview by Michael O'Leary, November 6, 2019. Unless otherwise noted, all further remarks from Hall are from this interview.

119 they wrote a memo: Brian Kahn, "Exxon Predicted 2019's Ominous CO2 Milestone in 1982," Gizmodo, May 14, 2019, https://earther.gizmodo.com/exxon-predicted-2019-s-ominous-co2-milestone-in-1982-1834748763.

119 "would warm the earth's surface": Ibid.

119 Meanwhile, global energy consumption: Jesse Barron, "How Big Business Is Hedging Against the Apocalypse," *New York Times*, April 11, 2019, https://www.nytimes.com/interactive/2019/04/11/magazine/climate-change-exxon-renewable-energy.html.

120 more than one thousand institutions: Yossi Cadan, Ahmed Mokgopo, and Clara Vondrich, "$11 Trillion and Counting," 350.org, https://financingthefuture.platform350.org/wp-content/uploads/sites/60/2019/09/FF_11Trillion-WEB.pdf.

120 That's up from $52 billion: Monica Tyler-Davies, "A New Fossil-Free Milestone: $11 Trillion Has Been Committed to Divest from Fossil Fuels," September 8, 2019, 350.org, https://350.org/11-trillion-divested/.

120 "I was twelve": Ilana Cohen, interview by Michael O'Leary, November 6, 2019. Unless otherwise noted, all further remarks from Cohen are from this interview.

120 "For someone growing up": Galen Hall, interview by Michael O'Leary, November 19, 2019. Unless otherwise noted, all further remarks from Hall are from this interview.

121 END HARVARD SUPPORT OF APARTHEID: Melissa C. Rodman and Yehong Zhu, "Calls for Divestment: A Retrospective," *Harvard*

Crimson, May 27, 2015, https://www.thecrimson.com/article/2015/5/27/divest-retrospective-reunion-1990/.

121 divestment was only partial: Rodman and Zhu, "Calls for Divestment."

122 "I've always had": Caleb Schwartz, interview by Michael O'Leary, October 30, 2019. Unless otherwise noted, all further remarks from Schwartz are from this interview.

122 "The endowment exists": Alexandra A. Chaidez, "Amid Student Calls for a Forum on Divestment, Bacow Remains Steadfast," *Harvard Crimson*, March 19, 2019, https://www.thecrimson.com/article/2019/3/19/bacow-reiterates-divestment-position/.

122 "has the financial resources": Harvard Management Company website, 2019, https://www.hmc.harvard.edu/about/.

123 "Just as the University": "Trustees Adopt New Investment Responsibility Framework for Stanford; University Commits $10 Million to Educational and Research Initiative," *Stanford News*, December 4, 2019, https://news.stanford.edu/2018/12/04/trustees-adopt-new-investment-responsibility-framework-stanford-university-commits-10-million-educational-research-initiative/.

123 Stanford's commitment: "Message from Marc Tessier-Lavigne and Persis Drell on Stanford's Commitment to Progress on Climate Change," *Stanford News*, June 5, 2017, https://news.stanford.edu/2017/06/05/stanfords-message-climate-change/; Drew Gilpin Faust and John L. Hennessy, "What Universities Can Do About Climate Change," *HuffPost*, September 23, 2014, updated November 23, 2014, https://www.huffpost.com/entry/post_b_5871214.

123 Harvard's operating expenses: Harvard University, *Financial Report, Fiscal Year 2018*, October 25, 2018, https://finance.harvard.edu/files/fad/files/harvard_annual_report_2018_final.pdf.

123 Its endowment at the time: John S. Rosenberg, "Harvard Endowment Increases 5.7 Percent to $39.2 Billion," *Harvard Magazine*, September 28, 2018, https://www.harvardmagazine.com/2018/09/harvard-endowment-39-2-billion-on-10-percent-return.

123 could buy a set of companies: Aswath Damodaran, NYU Stern School of Business, "Revenue Multiples by Sector (US)," January

2020, http://pages.stern.nyu.edu/~adamodar/New_Home_Page/datafile/psdata.html.

125 "a Quaker trust": "About Us," Joseph Rowntree Charitable Trust, January 20, 2020, https://www.jrct.org.uk/about-us.

125 "for your efforts": "Ibid.

125 "do not stand in isolation": "Recognition for Joseph Rowntree Charitable Trust's Responsible Investing," *Alliance*, September 14, 2019, https://www.alliancemagazine.org/blog/joseph-rowntree-charitable-trust-receive-recognition-for-responsible-investing/.

125 "We can't just go": Chaidez and Ryan, "Divest Protestors Pay Bacow a Morning Visit at Elmwood."

126 The public company sells: "NRG Energy," Wikipedia, October 16, 2019, https://en.wikipedia.org/wiki/NRG_Energy.

126 "The day is coming": David Crane, "NRG CEO: Here's How the Energy Industry Will Reinvent Itself," *GreenBiz*, March 27, 2014, https://www.greenbiz.com/blog/2014/03/27/david-crane-nrg-ceo-make-it-happen.

126 So Crane led NRG: Diane Cardwell and Alexandra Stevenson, "NRG, a Power Company Leaning Green, Faces Activist Challenge," *New York Times*, April 7, 2017, https://www.nytimes.com/2017/04/07/business/dealbook/nrg-elliott-management-climate.html.

126 Barry Smitherman: Ibid.

126 NRG released a plan: Unsurprisingly, given the state of corporate social responsibility doublespeak, the sustainability section of NRG's website is nevertheless bannered with the feel-good slogan "Creating change in the energy market by prioritizing sustainability." We'd hate to see what deprioritizing looks like. Tom DiChristopher, "Hedge Fund Titan Paul Singer Scores Big Win After NRG Energy Surges 25% in Single Day," CNBC, July 12, 2017, https://www.cnbc.com/2017/07/12/hedge-fund-titan-paul-singer-scored-a-big-win-with-this-energy-bet.html.

126 The board forced out Crane: Stephen Lacey, "David Crane Exits NRG with a Warning: 'There Is No Growth in Our Sector Outside of Clean Energy,'" Greentech Media, January 8, 2016, https://

www.greentechmedia.com/articles/read/david-crane-exits-nrg-with-a-warning.

127 it could speak for: Ed Crooks, "Activists Clash over Direction for NRG Energy," *Financial Times*, April 4, 2017, https://www.ft.com/content/89417ba2-1d3e-11e7-a454-ab04428977f9.

127 "there was no institutional investor support": David Crane, "Inside the Rise and Fall of NRG's Green Strategy," *GreenBiz*, July 17, 2017, https://www.greenbiz.com/article/inside-rise-and-fall-nrgs-green-strategy.

128 an unhappy shareholder: John Kay, *The Kay Review of UK Equity Markets and Long-Term Decision Making: Final Report* (London: Her Majesty's Government, July 2012), https://assets.publishing.service.gov.uk/government/uploads/system/uploads/attachment_data/file/253454/bis-12-917-kay-review-of-equity-markets-final-report.pdf, 21.

129 public companies have raised: Justin Fox and Jay W. Lorsch, "What Good Are Shareholders?," *Harvard Business Review*, July–August 2012, 50.

129 "Divestment, to date": Andrew Edgecliffe-Johnson and Billy Nauman, "Fossil Fuel Divestment Has 'Zero' Climate Impact, Says Bill Gates," *Financial Times*, September 17, 2019, https://www.ft.com/content/21009e1c-d8c9-11e9-8f9b-77216ebe1f17.

129 "divestment campaigns": Tyler Hansen and Robert Pollin, "Economics and Climate Justice Activism: Assessing the Fossil Fuel Divestment Movement," Political Economy Research Institute Working Paper no. 462, April 2018, https://www.peri.umass.edu/economists/robert-pollin/item/download/776_1ae306cb99a0ace3d577fe91ef68b0ab, abstract.

130 "Most efforts": C. J. Polychroniou, "Are Fossil Fuel Divestment Campaigns Working? A Conversation with Economist Robert Pollin," *Global Policy*, May 29, 2018, https://www.globalpolicyjournal.com/blog/29/05/2018/are-fossil-fuel-divestment-campaigns-working-conversation-economist-robert-pollin.

130 "If the aim of divestment": William MacAskill, "Does Divestment Work?," *The New Yorker*, October 20, 2015, https://www.newyorker.com/business/currency/does-divestment-work.

130 "It's near-impossible to prove": Rebecca Leber, "Divestment Won't Hurt Big Oil, and That's OK," *The New Republic*, May 20, 2015, https://newrepublic.com/article/121848/does-divestment-work.

131 porn accounts for 25 percent: "Internet Pornography by the Numbers; A Significant Threat to Society," Webroot, https://www.webroot.com/us/en/resources/tips-articles/internet-pornography-by-the-numbers. Figures as of September 25, 2019, "Top Websites Ranking," SimilarWeb, November 1, 2019, https://www.similarweb.com/top-websites/united-states.

133 In 2019, Monster Beverage: This and the following details are from Maria Gallagher and John Rotonti, "A Conversation on ESG and Sustainability with As You Sow CEO Andrew Behar," The Motley Fool, August 16, 2019, https://www.fool.com/investing/2019/07/25/esg-sustainability-as-you-sow-andrew-behar.aspx.

133 The SEC states that: Raj Gnanarajah and Gary Shorter, "Introduction to Financial Services: Corporate Governance," IF11221, U.S. Library of Congress, Congressional Research Service, May 21, 2019, https://crsreports.congress.gov/product/pdf/IF/IF11221, 2.

134 In 2016, this resulted: Norges Bank Investment Management, "Responsible Investment: Government Pension Fund Global," 2016, 42, https://www.nbim.no/contentassets/2c3377d07c5a4c4fbd442b345e7cfd67/government-pension-fund-global—responsible-investment-2016.pdf.

134 "calling for reports": Ross Kerber, "Exclusive: Fidelity May Back Climate Resolutions, a Milestone for Activists," Reuters, May 26, 2017, https://www.reuters.com/article/us-fidelity-climatechange/exclusive-fidelity-may-back-climate-resolutions-a-milestone-for-activists-idUSKBN18M110.

134 CalPERS manages $320 billion: CalPERS, "CalPERS' Climate Risk Reporting Proposal Passes at Occidental Petroleum," May 12, 2017, https://www.calpers.ca.gov/page/newsroom/calpers-news/2017/climate-risk-reporting-passes-occidental-petroleum.

134 require Occidental to explicitly account: Ibid.

134 The passage of this resolution: Ross Kerber, "BlackRock Switch Helps Pass 'Historic' Climate Measure at Occidental," Reuters,

May 12, 2017, https://www.reuters.com/article/us-blackrock-occidental-climate/blackrock-switch-helps-pass-historic-climate-measure-at-occidental-idUSKBN1882AA.

134 That same year: Steven Mufson, "Financial Firms Lead Shareholder Rebellion against ExxonMobil Climate Change Policies," *Washington Post*, May 31, 2017, https://www.washingtonpost.com/news/energy-environment/wp/2017/05/31/exxonmobil-is-trying-to-fend-off-a-shareholder-rebellion-over-climate-change/.

135 none has led to material change: Andy Behar and Danielle Fugere, *2020: A Clear Vision for Paris Compliant Shareholder Engagement* (Oakland, CA: As You Sow, September 6, 2018).

137 Three hundred eighty-eight thousand tons: Alan Rohn, "Napalm in Vietnam War," The Vietnam War, January 18, 2014, https://thevietnamwar.info/napalm-vietnam-war/; Sarah C. Haan, "Civil Rights and Shareholder Activism: SEC v. Medical Committee for Human Rights," *Washington and Lee Law Review* (forthcoming), October 18, 2019, updated November 22, 2019, accessed January 16, 2020, Washington & Lee Legal Studies Paper No. 2019-23, available at SSRN: https://ssrn.com/abstract=3472072.

137 However, on average: For recent shareholder proxy proposal sources, see "Shareholder-Sponsored Proxy Proposals," exhibit 1 in David F. Larcker, Brian Tayan, Vinay Trivedi, and Owen Wurzbacher, "Stakeholders and Shareholders: Are Executives Really 'Penny Wise and Pound Foolish' about ESG?," *Stanford Closer Look Series*, July 2, 2019, https://www.gsb.stanford.edu/sites/gsb/files/publication-pdf/cgri-closer-look-78-esg-programs.pdf, 4; Sullivan & Cromwell LLP, 2018 Proxy Season Review, July 12, 2018, https://www.sullcrom.com/2018-proxy-season-review.

137 The resolution at ExxonMobil: Mufson, "Financial Firms Lead Shareholder Rebellion."

138 Though BlackRock has publicly championed: Michael Holder, "'Get Ahead of These Risks': BlackRock Issues Climate Risk Warning to Investors," BusinessGreen, April 5, 2019, https://www.businessgreen.com/bg/news-analysis/3073716/get-ahead-of-these-risks-blackrock-issues-climate-risk-warning-to-investors.

138 it has supported 99 percent: Sierra Club, "Groups Deliver 129,000+ Signatures Demanding BlackRock Act on Climate," April 30,

2019, https://www.sierraclub.org/press-releases/2019/04/groups-deliver-129000-signatures-demanding-blackrock-act-climate.

138 A review of almost four thousand: Leo E. Strine, Jr., "Fiduciary Blind Spot: The Failure of Institutional Investors to Prevent the Illegitimate Use of Working Americans' Savings for Corporate Political Spending," Harvard Law School John M. Olin Center Discussion Paper no. 1022, December 20, 2018, https://pdfs.semanticscholar.org/9e02/a25917d61bacefbf679a1353b63887b8f684.pdf, 14.

138 "For decades": Ibid., 6.

138 For 90 percent of the companies: Ibid., 43.

138 the endowment is under: Harvard University, Corporation Committee on Shareholder Responsibility, "Annual Report 2017–2018," https://www.harvard.edu/sites/default/files/content/CCSR%20Annual%20Report%202018%20-%20FINAL-c.pdf.

139 But the two companies: Harvard Management Company, "Sustainable Investment Update," November 2019, http://www.hmc.harvard.edu/content/uploads/2019/11/Sustainable-Investing-Update-2019.pdf.

139 "When considering company strategy": Harvard University, Corporation Committee on Shareholder Responsibility, 15.

141 the phrase "climate change": Divest Harvard and Fossil Free Yale, "Opinion: We Disrupted the Harvard–Yale Game Because Our Schools Profit from Disaster," BuzzFeed News, November 26, 2019, https://www.buzzfeednews.com/article/divestharvardyale/opinion-why-we-disrupted-the-harvard-yale-game.

142 "That moment, when": Sam Gringlas, "Activists Disrupt Harvard-Yale Rivalry Game to Protest Climate Change," NPR, November 24, 2019, https://www.npr.org/2019/11/24/782427425/activists-disrupt-harvard-yale-rivalry-game-to-protest-climate-change.

CHAPTER 6: WE'RE GOING TO NEED A BIGGER BOAT

144 "twenty-six-year-old chain smoker": William F. Campbell, *Larkin Street Youth Services: The First 25 Years* (San Francisco: Larkin Street Youth Services, 2009), 10, https://larkinstreetyouth.org/wp-content/uploads/12_25-Year-History-Book-web.pdf.

144 Today, there are 1,300: Abhilash Mudaliar and Hannah Dithrich, "Sizing the Impact Investing Market," Global Impact Investing Network, April 1, 2019, https://thegiin.org/assets/Sizing%20the%20Impact%20Investing%20Market_webfile.pdf.

144 "Burning Man meets Wall Street": Jed Emerson, interview by Warren Valdmanis, September 11, 2019. Unless otherwise noted, all further remarks from Emerson are from this interview.

144 "have grown lazy": A version of his remarks was published as a blog post the following day: Jed Emerson, "Impact Investing Must Be Viewed as a Form of Capital Resistance," *Impact Investing,* SOCAP17, October 13, 2017, https://socialcapitalmarkets.net/2017/10/reflections-on-impact-resistance/.

146 "for purposes connected with": Pilar Palaciá nd Elisabetta Rurali, *Bellagio Center Villa Serbelloni: A Brief History,* trans. Paola Bianchi, 4, https://www.rockefellerfoundation.org/report/bellagio-center-villa-serbelloni-a-brief-history/.

146 The United Nations estimates: Scott Amyx, "Scaling Impact Investing to Trillions," *Forbes,* May 28, 2019, https://www.forbes.com/sites/forbesnycouncil/2019/05/28/scaling-impact-investing-to-trillions/; Mara Niculescu, "Impact Investment to Close the SDG Funding Gap," United Nations Development Programme, July 13, 2017, https://www.undp.org/content/undp/en/home/blog/2017/7/13/What-kind-of-blender-do-we-need-to-finance-the-SDGs-.html.

147 After all, the Fortune 500: "Fortune 500," *Fortune,* as of December 20, 2019, https://fortune.com/fortune500/.

147 Bugg-Levine coined the term: Saadia Madsbjerg, "Bringing Scale to the Impact Investing Industry," Rockefeller Foundation, August 15, 2018, https://www.rockefellerfoundation.org/blog/bringing-scale-impact-investing-industry/.

147 Pfund has led DBL to success: Nancy Pfund, "The Impact Generation," *HuffPost,* July 7, 2016, https://www.huffpost.com/entry/the-impact-generation_b_7744468.

147 investors in DBL's funds: Based on cofounder Mike Dorsey's claims on his LinkedIn page, https://www.linkedin.com/in/michael-dorsey-372a6710.

148 "I feel like I've spent": Andy Rosen, "Deval Patrick Revels in Role

Developing Bain's Social Impact Firm," *Boston Globe*, July 18, 2017, https://www.bostonglobe.com/business/2017/07/17/patrick-revels-role-developing-new-social-impact-bain-fund/hWtzrhYvIBPWVPL0UGKBqM/story.html.

148 "At the conclusion of our life": Antony Bugg-Levine and Jed Emerson, *Impact Investing: Transforming How We Make Money While Making a Difference* (San Francisco: Jossey-Bass/Wiley, 2011).

148 He did so on the principle: Joi Ito and Louis Kang, "Impact Investment Metrics and Their Limitations," Joi Ito, July 22, 2019, https://joi.ito.com/weblog/2019/07/22/measuring-impact.html.

150 Two years after the fund's launch: Leslie P. Norton, "TPG Cracks the Code for Impact Investing," *Barron's*, September 28, 2018, https://www.barrons.com/articles/tpg-cracks-the-code-for-impact-investing-1537567638.

150 "a lot of bad deals": Andrew Ross Sorkin, "A New Fund Seeks Both Financial and Social Returns," *New York Times*, December 19, 2016, https://www.nytimes.com/2016/12/19/business/dealbook/a-new-fund-seeks-both-financial-and-social-returns.html.

150 "just plunging ahead": Marc Gunther, "Hewlett Foundation's Leader Makes a Case Against Impact Investing," *Chronicle of Philanthropy*, January 8, 2019, https://www.philanthropy.com/article/Hewlett-Foundation-s-Leader/245394.

151 TPG would make an impact investment: Chris Addy, Maya Chorengel, Mariah Collins, and Michael Etzel, "Calculating the Value of Impact Investing," *Harvard Business Review*, January–February 2019, https://hbr.org/2019/01/calculating-the-value-of-impact-investing.

151 TPG calculated that: Ibid.

151 TPG then assigned: Ibid.

151 even the Department of Transportation report: US Department of Transportation, "Revised Departmental Guidance 2016: Treatment of the Value of Preventing Fatalities and Injuries in Preparing Economic Analyses," 2016, https://www.transportation.gov/sites/dot.gov/files/docs/2016%20Revised%20Value%20of%20a%20Statistical%20Life%20Guidance.pdf.

151 The Department of Agriculture: Dave Merrill, "No One Values

Your Life More Than the Federal Government," Bloomberg, October 19, 2017, https://www.bloomberg.com/graphics/2017-value-of-life/.

153 "The real trouble": Gilbert K. Chesterton, *Orthodoxy* (New York: Dodd, Mead & Company, 1936 [1908]), quoted in "From *Orthodoxy*," http://www.hjkeen.net/halqn/orthodx2.html.

153 "As we know": Ito and Kang, "Impact Investment Metrics."

155 At the beginning of 2019: Coca-Cola historical share price, January 2, 2019, retrieved from Coca-Cola, https://investors.coca-colacompany.com/stock-information/historical-data; shares outstanding data from Coca-Cola, Annual Report 2018, 71, https://investors.coca-colacompany.com/sec-filings/annual-reports/content/0000021344-19-000014/0000021344-19-000014.pdf; earnings per share data from Coca-Cola, Reconciliation of Non-GAAP Financial Measures for 2019 Consumer Analyst Group of New York (CAGNY) Conference, 7.

155 If you wanted to buy: In reality, if you tried to buy every share, you would have to pay significantly more than $45 per share, since there are many shareholders who would be unwilling to part with their shares for that amount. The incremental amount you'd have to pay over the current share price is known as the takeover premium, and it usually ranges from 20 to 40 percent.

156 PepsiCo's valuation multiple: PepsiCo historical share price, January 2, 2019, retrieved from PepsiCo, https://www.pepsico.com/investors/stock-information; earnings per share data from PepsiCo, PepsiCo Reports Fourth-Quarter and Full-Year 2018 Results, 1.

157 the present value: Alfred Rappaport, "The Economics of Short-Term Performance Obsession," *Financial Analysts Journal* 61, no. 3 (2005): 66.

157 For a typical company: John Kay, *The Kay Review of UK Equity Markets and Long-Term Decision Making: Final Report* (London: Her Majesty's Government, July 2012), 33.

158 The question is: This is a concept that the British economist John Kay describes as "obliquity." John Kay, *Obliquity: Why Our Goals Are Best Achieved Indirectly* (New York: Penguin, 2011).

158 Corporations have both: E. Merrick Dodd, Jr., "For Whom Are

Corporate Managers Trustees?," *Harvard Law Review* 45, no. 7 (May 1932): 1148.

158 we allow businesses to exist: Ibid., 1149.

158 "a sense of social responsibility": Ibid., 1160.

158 Berle argued that: A. A. Berle, Jr., "Corporate Powers as Powers in Trust," *Harvard Law Review* 44, no. 7 (May 1931): 1049.

158 "Either you have a system": Adolf A. Berle, Jr., "For Whom Corporate Managers Are Trustees: A Note," *Harvard Law Review* 45, no. 8 (June 1932): 1368, https://edisciplinas.usp.br/pluginfile.php/357201/mod_resource/content/0/Berle.%20For%20whom%20are%20corporate%20managers%20trustees%20-%20a%20note.%20Harvard%20Law%20Review%2C%20v.%2045%2C%20p.%201365%2C%201932.pdf.

159 The very term "corporate social responsibility": Howard R. Bowen, *Social Responsibilities of the Businessman* (New York: Harper & Brothers, 1953); republished in 2013 by University of Iowa Press with a foreword by Peter Geoffrey Bowen and an introduction by Jean-Pascal Gond.

159 93.5 percent of executives: Archie B. Carroll, "Corporate Social Responsibility: Evolution of a Definitional Concept," *Business & Society* 38, no. 3 (September 1999): 270, https://www.researchgate.net/publication/282441223_Corporate_social_responsibility_Evolution_of_a_definitional_construct.

159 "Today, most managements": Clarence Francis, as quoted in Bowen, *Social Responsibilities of the Businessman* (Iowa City: University of Iowa Press, 2013), 49.

159 "not consistent with *any* version": Robert N. Anthony, "The Trouble with Profit Maximization," *Harvard Business Review* 38, no. 6 (November–December 1960), as adapted in *The Firm as an Entity: Implications for Economics, Accounting and the Law*, ed. Yuri Biondi, Arnaldo Canziani, and Thierry Kirat (London: Routledge, 2007), 210.

160 "In both cases": Anthony, "The Trouble with Profit Maximization," 212.

160 "When we work": Quoted in Daniel Brown and Rakesh Khurana, "Leading Socially Responsible, Value-Creating Corporations," in *Corporate Stewardship: Achieving Sustainable Effectiveness*, eds. Susan

Albers Mohrman, James O'Toole, and Edward E. Lawler III (New York: Routledge, 2017), 81.

160 "On the face of it": Quoted in Brown and Khurana, "Leading Socially Responsible, Value-Creating Corporations," 79, from an interview with Welch by Francesco Guerrera, *Financial Times*, March 12, 2009.

160 He titled his book: Bill Walsh with Steve Jamison and Craig Walsh, *The Score Takes Care of Itself: My Philosophy of Leadership* (New York: Portfolio/Penguin, 2009).

162 Since the program's launch in 2006: Principles for Responsible Investment, brochure, January 28, 2019, https://www.unpri.org/download?ac=6303.

163 one study showing that: "Letter from JANA Partners & CalSTRS to Apple, Inc.," Harvard Law School Forum on Corporate Governance and Financial Regulation, January 19, 2018, https://corpgov.law.harvard.edu/2018/01/19/joint-shareholder-letter-to-apple-inc/. This is as opposed to those who spent less than an hour per day on devices.

163 the average American teen: Ibid.

163 "We believe there is": Ibid.

163 The writers recommended: Ibid.

163 Together, they owned: Robert G. Eccles, "Why an Activist Hedge Fund Cares Whether Apple's Devices Are Bad for Kids," *Harvard Business Review*, January 16, 2018, https://hbr.org/2018/01/why-an-activist-hedge-fund-cares-whether-apples-devices-are-bad-for-kids.

163 "The results were, frankly": Barry Rosenstein and Charles Penner, interview by Michael O'Leary, December 2, 2019. Unless otherwise noted, all further remarks from Rosenstein and Penner are from this interview.

163 Apple responded: Rene Ritchie, "Apple: New Parental Control Features Are Coming," iMore, January 8, 2018, https://www.imore.com/apple-new-parental-control-features-planned-future.

163 "digital wellbeing" features: Tim Bradshaw, "New Apple Software to Help Tackle Screen Addiction," *Financial Times*, June 4, 2018, https://www.ft.com/content/6f0b5b4e-6843-11e8-8cf3-0c230fa67aec.

163 a new impact investing effort at JANA: Svea Herbst-Bayliss, "Activist Investor Jana Hired Staff for New Socially Responsible Fund," Reuters, April 17, 2018, https://www.reuters.com/article/us-hedgefunds-jana/activist-investor-jana-hired-staff-for-new-socially-responsible-fund-idUSKBN1HO200.

164 He helped push: Nathan Vardi, "ValueAct Hedge Fund's Huge Microsoft Victory," *Forbes*, September 3, 2013, https://www.forbes.com/sites/nathanvardi/2013/09/03/valueact-hedge-funds-huge-microsoft-victory/.

164 annualized returns of 20 percent: Before fees; 15 percent after fees. Svea Herbst-Bayliss, "Ubben's Socially Conscious ValueAct Spring Fund Bets on Workplace Wonk," Reuters, February 27, 2019, https://www.reuters.com/article/us-hedgefunds-valueact/ubbens-socially-conscious-valueact-spring-fund-bets-on-workplace-wonk-idUSKCN1QG1G8.

165 "The ultimate goal": Jeff Ubben, interview by Michael O'Leary, November 11, 2019. Unless otherwise noted, all further remarks from Ubben are from this interview.

166 Ubben stepped down: "Watch CNBC's Exclusive Interview with ValueAct's Jeff Ubben," CNBC video, 9:04, from *Squawk Alley* episode aired by CNBC on April 16, 2019, https://www.cnbc.com/video/2019/04/16/watch-cnbcs-exclusive-interview-with-valueacts-jeff-ubben.html.

166 "I think this is": Cliff Robbins, interview by Michael O'Leary, November 25, 2019. Unless otherwise noted, all further remarks from Robbins are from this interview.

166 "I really believe": "Friendly Activist Cliff Robbins on ESG Investing," CNBC video, 4:30, from *Squawk Alley* episode aired by CNBC on April 17, 2018, https://www.cnbc.com/video/2018/04/17/friendly-activist-cliff-robbins-on-esg-investing.html.

166 "It makes me wonder": Kevin Stankiewicz, "Blue Harbour's Cliff Robbins: CEOs Will Eventually Tout Social Impact Score alongside Bond Rating," CNBC, September 19, 2019, https://www.cnbc.com/2019/09/19/blue-harbours-cliff-robbins-esg-investing-leads-to-higher-returns.html.

167 "The goals of equity markets": Kay, *The Kay Review of UK Equity Markets*, 14.

167 "Returns to beneficial owners": Ibid., 41.

168 two-thirds of Americans: Luigi Zingales, "Does Finance Benefit Society?," presidential address, American Finance Association Annual Meeting, Boston, MA, January 4, 2015, https://faculty.chicagobooth.edu/luigi.zingales/papers/research/finance.pdf, 2.

168 "When enterprise becomes": John Maynard Keynes, *The General Theory of Employment, Interest, and Money* (London: Macmillan, 1936), Book IV, chapter 12, available at http://cas2.umkc.edu/economics/people/facultypages/kregel/courses/econ645/winter2011/generaltheory.pdf.

168 deserved a Nobel Peace Prize: "Rich List 2018: Profiles 500 – 584 = ," *Times* (London), May 13, 2018, https://www.thetimes.co.uk/article/sunday-times-rich-list-2018-profiles-500-584-jvc7jqjzf.

168 the beginning of a revolution: Ronald Cohen, interview by Michael O'Leary, December 17, 2019. Unless otherwise noted, all further remarks from Cohen are from this interview.

168 "Governments are crucial": Sir Ronald Cohen, "An Impact Revolution with Sir Ronald Cohen," Stanford Graduate School of Business, presentation and panel discussion, February 25, 2019, https://youtu.be/EcXMzkRLB3M.

CHAPTER 7: DIVIDED AGAINST OURSELVES

171 And he has raised: Kate Clark, "Earnin Raises $125M to Help Workers Track and Cash Out Wages in Real Time," TechCrunch, December 20, 2018, https://techcrunch.com/2018/12/20/earnin-raises-125m-to-help-workers-track-and-cash-out-wages-in-real-time/.

171 Four out of five Americans: Emmie Martin, "The Government Shutdown Spotlights a Bigger Issue: 78% of US Workers Live Paycheck to Paycheck," CNBC, January 10, 2019, https://www.cnbc.com/2019/01/09/shutdown-highlights-that-4-in-5-us-workers-live-paycheck-to-paycheck.html.

171 Two in five: Sarah O'Brien, "Fed Survey Shows 40 Percent of Adults Still Can't Cover a $400 Emergency Expense," CNBC, May 23, 2018, https://www.cnbc.com/2018/05/22/fed-survey-40-percent-of-adults-cant-cover-400-emergency-expense.html.

171 they end up spending: "White Paper: On-Demand Payroll,"

Earnin, March 23, 2016, https://d3jlszua3q4cyr.cloudfront.net/content/web/EarninWhitePaper.pdf.

171 one in eight Netflix transactions: Ram Palaniappan, interview by Michael O'Leary, October 24, 2019. Unless otherwise noted, all remarks from Palaniappan are from this interview.

172 80 percent of users tip: Cyrus Farivar, "Millions Use Earnin to Get Cash Before Payday. Critics Say the App Is Taking Advantage of Them," NBC News, July 26, 2019, https://www.nbcnews.com/tech/internet/millions-use-earnin-get-cash-payday-critics-say-app-taking-n1034071.

172 Together, they voluntarily pay: Ibid.

172 than there are McDonald's: Jeannette N. Bennett, "Fast Cash and Payday Loans," Federal Reserve Bank of St. Louis, April 2019, https://research.stlouisfed.org/publications/page1-econ/2019/04/10/fast-cash-and-payday-loans.

172 The average interest rate: Megan Leonhardt, "This Map Shows the States Where Payday Loans Charge Nearly 700 Percent Interest," CNBC, August 3, 2018, https://www.cnbc.com/2018/08/03/states-with-the-highest-payday-loan-rates.html.

172 twenty-four times: Kelly Dilworth, "Average Credit Card Interest Rates: Week of Nov. 13, 2019," CreditCards.com, https://www.creditcards.com/credit-card-news/rate-report.php.

172 one hundred times: Yowana Wamala, "Average Auto Loan Interest Rates: 2019 Facts & Figures," ValuePenguin, September 3, 2019.

172 ends up paying $520: Jeannette N. Bennett, "Fast Cash and Payday Loans," Page One Economics, Federal Reserve Bank of St. Louis, April 2019, https://research.stlouisfed.org/publications/page1-econ/2019/04/10/fast-cash-and-payday-loans.

172 Eighty percent of payday loans: Ibid.

173 Another company, Elevate, charges: "About Us," Elevate, accessed December 21, 2019, https://www.elevate.com/company.html.

173 Elevate boasts that: Elevate home page, https://www.elevate.com/.

173 The Indian *Arthashastra* set: Raghuram Rajan, *The Third Pillar: How Markets and the State Leave the Community Behind* (New York: Penguin, 2019), 31: "The ceiling was 1¼ percent per month or 15 percent per year for ordinary loans to people. . . . It went up to 5 percent per month for ordinary commercial loans, 10 percent

per month for riskier commercial transactions that involved travel through forests, and 20 percent per month for trade by sea. . . . Thus, ancient India recognized a distinction between consumption loans and loans taken to fund profitable commerce, with lower ceilings on interest charged on the former. It also saw the need for the lender to receive a higher interest rate when the commercial enterprise was riskier."

173 A $5 tip on $100: Farivar, "Millions Use Earnin to Get Cash Before Payday."

173 still above many states' legal limits: Kevin Dugan, "Cash-Advance App Earnin Changes Its Tune amid NY Probe," *New York Post*, September 1, 2019, https://nypost.com/2019/09/01/cash-advance-app-earnin-changes-its-tune-amid-nys-probe/.

173 "Against economic tyranny": "June 27, 1936: Democratic National Convention," speech, University of Virginia Miller Center, https://millercenter.org/the-presidency/presidential-speeches/june-27-1936-democratic-national-convention.

174 bought back $4 trillion: Chuck Schumer and Bernie Sanders, "Limit Corporate Stock Buybacks," *New York Times*, February 3, 2019, https://www.nytimes.com/2019/02/03/opinion/chuck-schumer-bernie-sanders.html; William Lazonick, "Profits Without Prosperity," *Harvard Business Review*, September 2014, https://hbr.org/2014/09/profits-without-prosperity; John Aidan Byrne, "US Companies Spent $4T Buying Back Their Own Stock," *New York Post*, August 19, 2017, https://nypost.com/2017/08/19/us-companies-spent-4t-buying-back-their-own-stock/.

174 "so close to the line": Quote describing CIA activity in Jeff Stein, review of Michael V. Hayden, "Playing to the Edge: American Intelligence in the Age of Terror," *New York Times*, February 25, 2016, https://www.nytimes.com/2016/03/06/books/review/playing-to-the-edge-bymichael-v-hayden.html.

175 From 2008 through 2017: Schumer and Sanders, "Limit Corporate Stock Buybacks"; Lazonick, "Profits Without Prosperity."

175 that amounted to: David Kostin and Cole Hunter, "Debunking Buyback Myths," *Global Macro Research* (Goldman Sachs) 77 (April 11, 2019), 4, https://www.goldmansachs.com/insights/pages/top-of-mind/buyback-realities/report.pdf.

175 In 2018, 60 percent: Heather Slavkin Corso, "Petition for Rulemaking to Revise Rule 10b-18," Harvard Law School Forum on Corporate Governance, July 18, 2019, https://corpgov.law.harvard.edu/2019/07/18/petition-for-rulemaking-to-revise-rule-10b-18/.

175 Until 1982, share buybacks: Lazonick, "Profits Without Prosperity."

175 then John Shad: Lazonick, "Profits Without Prosperity."

175 "retain-and-reinvest": William Lazonick, "Stock Buybacks: From Retain-and-Reinvest to Downsize-and-Distribute," Center for Effective Public Management at Brookings, April 2015, 1, https://www.brookings.edu/wp-content/uploads/2016/06/lazonick.pdf.

176 In 2018, Boeing disbursed: "Fourth-Quarter 2018 Performance Review and 2019 Guidance," Boeing, January 30, 2019, https://s2.q4cdn.com/661678649/files/doc_financials/quarterly/2018/q4/4Q18-Presentation.pdf.

176 Boeing would divide: Boeing, *Annual Report 2018*, February 8, 2019, https://s2.q4cdn.com/661678649/files/doc_financials/annual/2019/Boeing-2018AR-Final.pdf, 73.

176 the Securities and Exchange Commission studied: Robert J. Jackson Jr., "Stock Buybacks and Corporate Cashouts," speech, June 11, 2018, https://www.sec.gov/news/speech/speech-jackson-061118#_ftnref22.

177 Cisco Systems spent $129 billion: Sheelah Kolhatkar, "The Economist Who Put Stock Buybacks in Washington's Crosshairs," *The New Yorker*, June 20, 2019, https://www.newyorker.com/business/currency/the-economist-who-put-stock-buybacks-in-washingtons-crosshairs.

177 Lehman Brothers bought back: William Lazonick, "Everyone Is Paying Price for Share Buy-Backs," *Financial Times*, September 25, 2008, https://www.ft.com/content/e75440f6-8b0e-11dd-b634-0000779fd18c.

177 "At a time of huge": Schumer and Sanders, "Limit Corporate Stock Buybacks."

177 Their bill would ban buybacks: Ibid.

177 As evidence, he shows: Marco Rubio, "American Investment in the 21st Century: Project for Strong Labor Markets and National

Development," May 15, 2019, https://www.rubio.senate.gov/public/_cache/files/9f25139a-6039-465a-9cf1-feb5567aebb7/4526E9620A9A7DB74267ABEA5881022F.5.15.2019.-final-project-report-american-investment.pdf.

177 proposed amending the tax code: Julius Krein, "Share Buybacks and the Contradictions of 'Shareholder Capitalism,'" *American Affairs,* December 13, 2018, https://americanaffairsjournal.org/2018/12/share-buybacks-and-the-contradictions-of-shareholder-capitalism/.

177 Senator Kamala Harris seeks: Li Zhou, "Kamala Harris's Plan to Close the Gender Wage Gap, Explained," Vox, May 21, 2019, https://www.vox.com/2019/5/21/18632793/kamala-harris-gender-pay-gap-fines-iceland.

178 revoke the corporate charters: "Accountable Capitalism Act," on Elizabeth Warren's Senate website, accessed January 20, 2020, https://www.warren.senate.gov/imo/media/doc/Accountable%20Capitalism%20Act%20One-Pager.pdf.

178 India passed a law: India's Companies Act of 2013, Section 135.

178 Firms that spent less: Dhammika Dharmapalaa and Vikramaditya Khannab, "The Impact of Mandated Corporate Social Responsibility: Evidence from India's Companies Act of 2013," *International Review of Law and Economics* 56 (December 2018), https://www.sciencedirect.com/science/article/abs/pii/S0144818818301182.

178 after the law was passed: Oliver Balch, "Indian Law Requires Companies to Give 2% of Profits to Charities: Is It Working?," *Guardian,* April 5, 2016, https://www.theguardian.com/sustainable-business/2016/apr/05/india-csr-law-requires-companies-profits-to-charity-is-it-working.

179 Ovitz had cofounded: "Ovitz Keeps $140m Disney Payout," CNN, August 10, 2005, http://edition.cnn.com/2005/BUSINESS/08/10/disney.ovitz/.

179 "Finally I settled down": Kim Masters, "The Epic Disney Blow-up of 1994: Eisner, Katzenberg and Ovitz 20 Years Later," *Hollywood Reporter,* April 9, 2014, https://www.hollywoodreporter.com/features/epic-disney-blow-up-1994-694476.

179 His severance package: Laura M. Holson, "Ruling Upholds

Disney's Payment in Firing of Ovitz," *New York Times*, August 10, 2005, https://www.nytimes.com/2005/08/10/business/media/ruling-upholds-disneys-payment-in-firing-of-ovitz.html.

179 "should not serve": Jill Goldsmith and Janet Shprintz, "Eisner Gets the Last Laugh; Judge Rules Ovitz Dough Was No Crime," *Variety*, August 9, 2005, https://variety.com/2005/biz/news/eisner-gets-the-last-laugh-1117927213/.

179 CEOs' pay has increased: Lawrence Mishel and Julia Wolfe, "CEO Compensation Has Grown 940% Since 1978," Economic Policy Institute, August 14, 2019, https://www.epi.org/publication/ceo-compensation-2018/.

179 The ratio of executives' pay: Ibid.

180 it would have reached $23: Leo E. Strine, Jr., "Toward a True Corporate Republic: A Traditionalist Response to Bebchuk's Solution for Improving Corporate America," *Harvard Law Review* 119, no. 6 (April 2006): 1767.

180 versus the $7.25: Alison Doyle, "2020 Federal and State Minimum Wage Rates," The Balance Careers, December 12, 2019, https://www.thebalancecareers.com/2018-19-federal-state-minimum-wage-rates-2061043.

180 When Bob Iger, Disney's current CEO: Ainsley Harris, "Disney CEO Bob Iger's Compensation Is 'Insane,' Says Abigail Disney," *Fast Company*, April 19, 2019, https://www.fastcompany.com/90333082/disney-ceo-bob-igers-compensation-is-insane-says-abigail-disney.

180 As far back as 1934: Kevin J. Murphy and Michael C. Jensen, "The Politics of Pay: The Unintended Consequences of Regulating Executive Compensation," USC Law Legal Studies Paper no. 18-8, April 20, 2018, 30.

180 90 percent believed that: Jess Whittlestone, "Do You Think You're Better Than Average?," 80,000 Hours, November 27, 2012, https://80000hours.org/2012/11/do-you-think-you-re-better-than-average/.

181 Nineteen out of twenty companies: Mike Kesner, Tara Tays, and Ed Sim, "CEO Pay Ratio: Leading Indicators of Broader Human Resource Matters?," Harvard Law School Forum on Corporate Governance, July 21, 2018, https://corpgov.law.harvard.

edu/2019/07/21/ceo-pay-ratio-leading-indicators-of-broader-human-resource-matters/.

181 Before the 1970s: Anat R. Admati, "A Skeptical View of Financialized Corporate Governance," *Journal of Economic Perspectives* 31, no. 3 (Summer 2017): 133.

181 the average CEO pay: John Roe and Kosmas Papadopoulos, "2019 U.S. Executive Compensation Trends," Harvard Law School Forum on Corporate Governance (blog post), April 16, 2019, https://corpgov.law.harvard.edu/2019/04/16/2019-u-s-executive-compensation-trends/.

181 Over two-thirds was in stock: Ibid.

182 At least once every three years: US Securities and Exchange Commission, "SEC Adopts Rules for Say-on-Pay and Golden Parachute Compensation as Required Under Dodd-Frank Act," press release, January 25, 2011, https://www.sec.gov/news/press/2011/2011-25.htm.

182 only 1.6 percent of firms: Murphy and Jensen, "The Politics of Pay," 2.

182 Most companies receive: Paul H. Edelman, Randall S. Thomas, and Robert B. Thompson, "Shareholder Voting in an Age of Intermediary Capitalism," *Southern California Law Review* 87, no. 6 (September 2014): 1428; Terry Newth and Dean Chaffee, "Ten Years of Say-on-Pay Data," Harvard Law School Forum on Corporate Governance, June 9, 2019, https://corpgov.law.harvard.edu/2019/06/09/ten-years-of-say-on-pay-data/.

182 Despite the dissenting voices: Jill Disis, "Disney Shareholders Narrowly Approve CEO Bob Iger's Pay Package," CNN, March 7, 2019, https://www.cnn.com/2019/03/07/media/disney-bob-iger-compensation/index.html.

183 spent 2.4 percent of GDP on R&D: Organisation for Economic Co-operation and Development, "Gross Domestic Spending on R&D (indicator)," doi: 10.1787/d8b068b4-en, accessed February 21, 2020, https://data.oecd.org/rd/gross-domestic-spending-on-r-d.htm.

183 While the federal government spent: Economic Innovation Group, "Dynamism in Retreat: Consequences for Regions, Markets, and Workers," February 2017, https://eig.org/wp-content/uploads/2017/07/Dynamism-in-Retreat-A.pdf.

183 New private investment: US Bureau of Economic Analysis, "Shares of Gross Domestic Product: Gross Private Domestic Investment (A006RE1Q156NBEA)," Federal Reserve Bank of St. Louis, February 21, 2020, https://fred.stlouisfed.org/series/A006RE1Q156NBEA.

184 keeping real interest rates: Robin Harding, "Profoundly Low Interest Rates Are Here to Stay," *Financial Times*, July 30, 2019, https://www.ft.com/content/84a1b13c-b2a3-11e9-8cb2-799a3a8cf37b.

184 A report by Goldman Sachs: David Kostin and Cole Hunter, "Debunking Buyback Myths," *Global Macro Research* (Goldman Sachs) 77 (April 11, 2019), 4, https://www.goldmansachs.com/insights/pages/top-of-mind/buyback-realities/report.pdf.

184 higher than average: Ibid.

184 sell their own shares: Jackson, "Stock Buybacks and Corporate Cashouts."

185 Within months, it would lose: Andrew Tangel, Andy Pasztor, and Mark Maremont, "The Four-Second Catastrophe: How Boeing Doomed the 737 MAX," *Wall Street Journal*, August 16, 2019, https://www.wsj.com/articles/the-four-second-catastrophe-how-boeing-doomed-the-737-max-11565966629; Boeing historical share price, March 1, 2019–August 1, 2019, retrieved from Boeing, https://investors.boeing.com/investors/stock-information/default.aspx.

185 By the summer: Patrick Thomas and Austen Hufford, "Boeing's Plane Deliveries Tumble as 737 MAX Jet Stays Grounded," *Wall Street Journal*, August 13, 2019, https://www.wsj.com/articles/boeing-plane-deliveries-tumble-so-far-in-2019-11565714502.

185 For its part: Peggy Hollinger, "Boeing Targets Share Buybacks to Cover Costs from 737 Max Crisis," *Financial Times*, October 23, 2019, https://www.ft.com/content/4549b164-f589-11e9-a79c-bc9acae3b654.

185 "We've also taken action": Ibid.

185 "It concerns us": Lazonick, "Profits Without Prosperity."

186 defending high drug prices: Jon Gingerich, "Lobbying Topped $3.4 Billion in 2018," O'Dwyer's, January 28, 2019, https://www.odwyerpr.com/story/public/11951/2019-01-28/lobbying-topped

-34-billion-2018.html; Susan Scutti, "Big Pharma Spends Record Millions on Lobbying Amid Pressure to Lower Drug Prices," CNN, January 24, 2019, https://www.cnn.com/2019/01/23/health/phrma-lobbying-costs-bn/index.html.

186 the first federal campaign finance law: Naomi R. Lamoreaux and William J. Novak, eds., *Corporations and American Democracy* (Cambridge, MA: Harvard University Press, 2017), 37.

187 if it ended up hurting Mylan: Edelman et al., "Shareholder Voting in an Age of Intermediary Capitalism," 1405.

187 Today, no legal duty binds: "The Modern Dilemma Balancing Short- and Long-Term Business Pressures," World Economic Forum, March 2019, http://www3.weforum.org/docs/WEF_Modern_Dilemma_Report_2019.pdf, 9.

188 introduced a Stewardship Code for investors: Owen Walker, "UK Stewardship Code in Line for Significant Overhaul," *Financial Times,* January 5, 2019, https://www.ft.com/content/e581e639-0ab9-3fbe-aed6-39f7a9b20734.

188 "as the responsible allocation": Financial Reporting Council, "UK Stewardship Code," October 24, 2019, https://www.frc.org.uk/investors/uk-stewardship-code.

188 "making investors' stewardship efforts": Richard M. Brand and Joanna Valentine, "2020 Update to the UK Stewardship Code," *National Law Review,* November 5, 2019, https://www.natlawreview.com/article/2020-update-to-uk-stewardship-code.

188 "systematically integrate stewardship": Financial Reporting Council, "The UK Stewardship Code 2020," https://www.frc.org.uk/getattachment/5aae591d-d9d3-4cf4-814a-d14e156a1d87/Stewardship-Code_Final2.pdf, 15.

188 "Signatories are now required": Louise Barber, "Corporate Governance: Revised UK Stewardship Code," Down the Wire, Squire Patton Boggs, October 28, 2019, https://www.downthewireblog.com/2019/10/corporate-governance-revised-uk-stewardship-code/.

188 "If the code remains": John Kingman, *Independent Review of the Financial Reporting Council* (London, UK: APS Group, December 2018), https://assets.publishing.service.gov.uk/government/uploads/system/uploads/attachment_data/file/767387/frc-independent-review-final-report.pdf, 46.

190 The government gave: Greenspan and Wooldridge, *Capitalism in America,* 113.

190 "Capitalism in America": Bhu Srinivasan, *Americana: A 400-Year History of American Capitalism* (New York: Penguin, 2017), 485.

191 the sulfur dioxide emitted: Richard Conniff, "The Political History of Cap and Trade," *Smithsonian Magazine,* August 2009, https://www.smithsonianmag.com/science-nature/the-political-history-of-cap-and-trade-34711212/.

191 Even though electricity generation: Richard Schmalensee and Robert N. Stavins, "Lessons Learned from Three Decades of Experience with Cap and Trade," *Review of Environmental Economics and Policy* 11, no. 1 (Winter 2017): 59–79, https://academic.oup.com/reep/article/11/1/59/3066276.

191 generating an estimated $122 billion: Conniff, "The Political History of Cap and Trade."

192 "Walmart pays its employees so little": "Walmart on Tax Day: How Taxpayers Subsidize America's Biggest Employer and Richest Family," Americans for Tax Fairness, April 2014, https://americansfortaxfairness.org/files/Walmart-on-Tax-Day-Americans-for-Tax-Fairness-1.pdf, 3.

192 "Just as we should globalize": Leo E. Strine, Jr., "The Delaware Way: How We Do Corporate Law and Some of the New Challenges We (and Europe) Face," *Delaware Journal of Corporate Law* 30 (2005): 696.

192 though private companies fund: John F. Sargent, Jr., *Federal Research and Development (R&D) Funding: FY2019,* Congressional Research Service, October 4, 2018, https://fas.org/sgp/crs/misc/R45150.pdf.

193 federally funded R&D spending: Economic Innovation Group, *Dynamism in Retreat: Consequences for Regions, Markets, and Workers,* February 2017, https://eig.org/wp-content/uploads/2017/07/Dynamism-in-Retreat-A.pdf.

193 with China close behind: John F. Sargent Jr., "Global Research and Development Expenditures," Congressional Research Service, updated June 27, 2018, 1, 2, https://crsreports.congress.gov/product/pdf/R/R44283/9.

194 "the North invested": Greenspan and Wooldridge, *Capitalism in America*, 80.

194 As a result, 93 percent: Ibid., 70.

194 Well into the twentieth century: Ibid., 81.

194 Nearly two in five Americans: Travis Hornsby, "How Many Americans Know What Compound Interest Is?," Millennial Moola, June 7, 2016, https://millennialmoola.com/2016/06/07/compound-interest-unknown/.

195 "An overwhelming majority": Luigi Zingales, "Does Finance Benefit Society?," presidential address, American Finance Association Annual Meeting, Boston, MA, January 4, 2015, https://faculty.chicagobooth.edu/luigi.zingales/papers/research/finance.pdf, 14.

195 seventeen US states: Cheryl R. Cooper, "An Overview of Consumer Finance and Policy Issues," Congressional Research Service, July 12, 2019, https://crsreports.congress.gov/product/pdf/R/R45813.

195 When other states have tried: National Consumer Law Center, "Issue Brief: Stop Payday Lenders' Rent-A-Bank Schemes," November 2019, https://www.nclc.org/images/Rent-a-bank-one-pager.pdf.

195 the proportion of US GDP spent: Greenspan and Wooldridge, *Capitalism in America*, 192.

195 Now nearly three out of four loans: Pew Charitable Trusts, "Trial, Error, and Success in Colorado's Payday Lending Reforms," December 2014, https://www.pewtrusts.org/~/media/assets/2014/12/pew_co_payday_law_comparison_dec2014.pdf.

195 The interest rate is 115 percent: Ibid.

195 borrowers spent 42 percent less: Ibid.

196 But that all happened: Zingales, "Does Finance Benefit Society?," 27.

196 "The law's transparent pricing": Pew Charitable Trusts, "Trial, Error, and Success in Colorado's Payday Lending Reforms," 6.

197 "What fed the industry's growth": Nicholas Confessore, "Mick Mulvaney's Master Class in Destroying a Bureaucracy from Within," *New York Times Magazine*, April 16, 2019, https://www

.nytimes.com/2019/04/16/magazine/consumer-financial-protection-bureau-trump.html.

197 seven of ten borrowers: Bennett, "Fast Cash and Payday Loans."

197 effective government is no substitute: Or responsible consumers—as we'll see in chapter 9.

197 some of capitalism's greatest beneficiaries: "Capitalism in Crisis: U.S. Billionaires Worry About the Survival of the System That Made Them Rich," *Washington Post*, April 20, 2019, https://www.washingtonpost.com/politics/capitalism-in-crisis-us-billionaires-worry-about-the-survival-of-the-system-that-made-them-rich/2019/04/20/3e06ef90-5ed8-11e9-bfad-36a7eb36cb60_story.html.

197 JPMorgan CEO Jamie Dimon: Jamie Dimon, "Chairman & CEO Letter to Shareholders," JPMorgan Chase & Co., September 15, 2018, https://reports.jpmorganchase.com/investor-relations/2018/ar-ceo-letters.htm.

197 hedge fund billionaire Ray Dalio: Ray Dalio, "Why and How Capitalism Needs to Be Reformed," Economic Principles, April 5, 2019, https://economicprinciples.org/Why-and-How-Capitalism-Needs-To-Be-Reformed/.

198 "The status quo is unacceptable": Andrew Ross Sorkin, "Walmart's C.E.O. Steps into the Gun Debate. Other C.E.O.s Should Follow," *New York Times*, September 5, 2019, https://www.nytimes.com/2019/09/03/business/walmart-ceo-guns.html.

198 "We take seriously": Walmart, "Statement on Firearms Policy," February 28, 2018, https://news.walmart.com/2018/02/28/walmart-statement-on-firearms-policy.

198 "'Watching those kids'": David Gelles, "The C.E.O. Taking On the Gun Lobby," *New York Times*, October 25, 2019, https://www.nytimes.com/2019/10/25/business/ed-stack-dicks-sporting-goods-corner-office.html.

199 Vanguard estimated that: Mark J. Roe, "Corporate Short-Termism—In the Boardroom and in the Courtroom," *The Business Lawyer* 68, no. 4 (August 2013): 1002.

199 "molecules of U.S. freedom": Ellie Kaufman, "U.S. Energy Officials Hail 'Freedom Gas,' 'Molecules of Freedom,'" CNN, May 29, 2019, https://www.cnn.com/2019/05/29/politics/doe-freedom-gas-natural-gas/index.html.

199 Whereas in 1960, 75 percent: Greenspan and Wooldridge, *Capitalism in America*, 305.

199 64 percent of Americans say: Marisa Fernandez, "By the Numbers: The Rise of 'Belief-Driven' Buyers," Axios, October 28, 2018, https://www.axios.com/belief-driven-activist-brands-nike-kaepernick-edelman-e4680a8e-7aa5-43e1-84de-2874929c868e.html.

199 "Business has to pick up": Alana Semuels, "'Rampant Consumerism Is Not Attractive.' Patagonia Is Climbing to the Top—and Reimagining Capitalism along the Way," *Time*, September 23, 2019, https://time.com/5684011/patagonia/.

CHAPTER 8: THE $23 TRILLION SOLUTION

202 Tylenol accounted for: Thomas Moore, "The Fight to Save Tylenol," *Fortune*, November 29, 1982, https://fortune.com/2012/10/07/the-fight-to-save-tylenol-fortune-1982/.

202 "There were many people": "Tylenol and the Legacy of J&J's James Burke," Knowledge@Wharton, October 2, 2012, https://knowledge.wharton.upenn.edu/article/tylenol-and-the-legacy-of-jjs-james-burke/.

202 "The Tylenol episode": Clyde Haberman, "How an Unsolved Mystery Changed the Way We Take Pills," *New York Times*, September 16, 2018, https://www.nytimes.com/2018/09/16/us/tylenol-acetaminophen-deaths.html.

205 "We believe our first responsibility": "Our Credo," Johnson & Johnson, https://www.jnj.com/credo/.

206 "industry only has the right": James O'Toole, *The Enlightened Capitalists: Cautionary Tales of Business Pioneers Who Tried to Do Well by Doing Good* (New York: Harper Business, 2019), 150.

207 whereas it would cost: Christopher Leggett, "The Ford Pinto Case: The Valuation of Life as It Applies to the Negligence-Efficiency Argument," Wake Forest University School of Law, Spring 1999, https://users.wfu.edu/palmitar/Law&Valuation/Papers/1999/Leggett-pinto.html.

207 recalling 1.4 million pintos: John Pearley Huffman, "5 Most Notorious Recalls of All Time," *Popular Mechanics*, February 12, 2010, https://www.popularmechanics.com/cars/g261/4345725/.

208 It soon reclaimed: D. Sirisha, "Johnson & Johnson's Tylenol Controversies," in *The Global Corporation: Sustainable, Effective and Ethical Practices: A Case Book*, ed. Laura P. Hartman and Patricia H. Werhane (New York: Routledge, 2009), 19.

208 In President Clinton's remarks: "James Burke Acceptance Presidential Medal of Freedom," C-SPAN video, 5:15, from the Presidential Medals of Freedom Ceremony televised by C-SPAN on August 9, 2000, https://www.c-span.org/video/?c4483558/user-clip-james-burke-acceptance-presidential-medal-freedom&start=486.

208 "Why does Etsy exist?": Josh Silverman, interview by Michael O'Leary, November 4, 2019. Unless otherwise noted, all further remarks from Silverman are from this interview.

208 The former chief executive of Skype: Maria Gallagher, "ESG Investing: Is Etsy a Responsible Investment?," The Motley Fool, October 8, 2019, https://www.fool.com/investing/2019/10/08/esg-investing-is-etsy-a-responsible-investment.aspx; David Gelles, "Inside the Revolution at Etsy," *New York Times*, November 25, 2017, https://www.nytimes.com/2017/11/25/business/etsy-josh-silverman.html.

208 "his enthusiasm for Etsy": Ibid.

208 "I was patient": Gelles, "Inside the Revolution at Etsy."

208 In describing Etsy's headquarters: Michelle Higgins, "Jehovah's Witnesses' Brooklyn Headquarters for Sale," *New York Times*, January 29, 2016, https://www.nytimes.com/2016/01/31/realestate/jehovahs-witnesses-brooklyn-headquarters-for-sale.html.

209 "He's a bald": Phil Wahba, "How Etsy Crafted an E-Commerce Comeback," *Fortune*, July 25, 2019, https://fortune.com/2019/07/25/etsy-ecommerce-growth-strategies/.

209 The day before Silverman: Anthony Ha, "Etsy Names Josh Silverman as Its New CEO," TechCrunch, May 2, 2017, https://techcrunch.com/2017/05/02/etsy-new-ceo/.

209 It was the first big layoff: Gelles, "Inside the Revolution at Etsy."

209 Silverman fired another 140: Ibid.

209 "sold $2.39 billion worth": Amy Larocca, "Etsy Wants to Crochet Its Cake, and Eat It Too," *The Cut*, April 5, 2016, https://www.thecut.com/2016/04/etsy-capitalism-c-v-r.html.

209 had remained unprofitable: Etsy, *Unlocking Opportunity: 2018 Annual Report*, https://s22.q4cdn.com/941741262/files/doc_financials/annual/2018-Annual-Report-(1).pdf.

209 Amazon launched: Hiroko Tabuchi, "Amazon Challenges Etsy with Strictly Handmade Marketplace," *New York Times*, October 8, 2015, https://www.nytimes.com/2015/10/08/business/amazon-challenges-etsy-with-strictly-handmade-marketplace.html.

209 closed the day: Sarah Buhr and Alex Wilhelm, "Etsy Closes Up 86 Percent on First Day of Trading," TechCrunch, April 16, 2015, https://techcrunch.com/2015/04/16/etsy-stock-surges-86-percent-at-close-of-first-day-of-trading-to-30-per-share/.

209 Etsy shares had lost: Etsy historical share price, April 16, 2015–February 10, 2016, retrieved from Etsy, https://investors.etsy.com/stock-info/stock-info/default.aspx.

209 According to *Bloomberg Businessweek*: Max Chafkin and Jing Cao, "The Barbarians Are at Etsy's Hand-Hewn, Responsibly Sourced Gates," *Bloomberg Businessweek*, May 18, 2017, https://www.bloomberg.com/news/features/2017-05-18/the-barbarians-are-at-etsy-s-hand-hewn-responsibly-sourced-gates.

210 "I'm not crying": Chafkin and Cao, "The Barbarians Are at Etsy's Hand-Hewn, Responsibly Sourced Gates."

210 "[The] corporation is": Quoted in Thomas P. Byrne, "False Profits: Reviving the Corporation's Public Purpose," *UCLA Law Review* 57 (2010): 29–30.

210 From 1790 to 1860: Ralph Gomory and Richard Sylla, "The American Corporation," in "American Democracy & the Common Good," special issue, *Daedalus* 142, no. 2 (Spring 2013): 104.

210 "it would be": The Granite Railway Company, *A History of the Origin and Development of the Granite Railway at Quincy, Massachusetts* (Quincy, MA: Thomas Crane Library, 2001 [1926]), http://thomascranelibrary.org/legacy/railway/railway.htm.

211 "employment of a large sum": Ibid.

211 "Associated individuals [want]": Quoted in Byrne, "False Profits," 44.

211 "A corporation . . . may be": Quoted in David Ciepley, "Beyond Public and Private: Toward a Political Theory of the Corporation," *American Political Science Review* 107, no. 1 (February 2013): 142.

211 "added the fuel": Abraham Lincoln, *The Collected Works of Abraham Lincoln*, Volume 3, edited by Roy P. Basler (New Brunswick, NJ: Rutgers University Press, 1953), available at https://quod.lib.umich.edu/l/lincoln/lincoln3/1:87?rgn=div1;view=fulltext.

211 maybe it shouldn't be: Maybe it represents a good opportunity for social impact bonds, per chapter 7.

211 Charters were gradually replaced: Byrne, "False Profits," 32.

211 general incorporation codes put an end: Ibid., 31.

212 Eighty-seven percent of the sellers: Etsy, *Unlocking Opportunity: 2018 Annual Report.*

213 targeted initiatives such as statewide sales: Wahba, "How Etsy Crafted an E-Commerce Comeback."

213 He cut half the initiatives: Ibid.

213 Gross merchandise sales surged: Etsy, *Unlocking Opportunity: 2018 Annual Report.*

215 The company calculated: Emily Dreyfuss, "Etsy Crafts a Plan for Carbon-Neutral Online Shopping," *Wired*, February 27, 2019, https://www.wired.com/story/etsy-carbon-neutral-online-shopping/.

215 it became the first: Josh Silverman, "Etsy Becomes the First Global eCommerce Company to Completely Offset Carbon Emissions from Shipping," Etsy News, February 26, 2019, https://blog.etsy.com/news/2019/on-etsy-every-purchase-makes-a-positive-impact/.

216 zero-waste operation: Ibid.

216 set the goals of doubling: Etsy, *Unlocking Opportunity: 2018 Annual Report.*

217 Among the board members: Lawrence G. McDonald, *A Colossal Failure of Common Sense: The Inside Story of the Collapse of Lehman Brothers* (New York: Crown Business, 2009), chap. 9.

217 Board members' salaries: Paul Ausick, "25 Companies That Pay Their Board of Directors a Shocking Amount," *USA Today*, December 14, 2018, https://www.usatoday.com/story/money/business/2018/12/14/how-much-do-corporate-boards-pay-companies-highest-compensation/38637377/.

217 that's less than five hours: Dannetta English Bland, "Business Chemistry: A Path to a More Effective Board Composition,"

Harvard Law School Forum on Corporate Governance, June 19, 2019, https://corpgov.law.harvard.edu/2019/06/19/business-chemistry-a-path-to-a-more-effective-board-composition/.

217 Some board members provide: Siri Terjesen, "Are Corporate Boards Doing Their Jobs?," *Industrial Equipment News*, June 24, 2019, https://www.ien.com/operations/news/21074626/are-corporate-boards-doing-their-jobs.

218 marketing Risperdal "off label": *United States of America ex rel. Victoria Starr v. Janssen Pharmaceutical Products, L.P.*, Civil Action No. 04-cv-1529, E.D. Pa., November 4, 2013, https://highline.huffingtonpost.com/miracleindustry/americas-most-admired-lawbreaker/assets/documents/3/janssen-call-notes.pdf?build=02281049.

218 "it can cause boys": Ibid; Nicholas Kristof, "When Crime Pays: J&J's Drug Risperdal," *New York Times*, September 17, 2015, https://www.nytimes.com/2015/09/17/opinion/nicholas-kristof-when-crime-pays-jjs-drug-risperdal.html.

218 Eventually, it was forced: Kristof, "When Crime Pays."

219 responded with one voice: Steven Brill, "The New CEO," part 10 of "America's Most Admired Lawbreaker," *Huffington Post*, Fall 2015, 15-part serialized piece on Risperdal, available at https://highline.huffingtonpost.com/miracleindustry/americas-most-admired-lawbreaker/chapter-10.html.

219 executives had been aware: Roni Caryn Rabin and Tiffany Hsu, "Johnson & Johnson Feared Baby Powder's Possible Asbestos Link for Years," *New York Times*, December 14, 2018, https://www.nytimes.com/2018/12/14/business/baby-powder-asbestos-johnson-johnson.html.

220 "If you look for trade-offs": Karen Christensen, "The Co-Founder of Whole Foods Makes a Case for 'Conscious Capitalism,'" *Globe and Mail*, October 29, 2013, updated May 11, 2018, https://www.theglobeandmail.com/report-on-business/careers/careers-leadership/the-co-founder-of-whole-foods-makes-a-case-for-conscious-capitalism/article14970731/.

220 "Etsy had the potential": Gelles, "Inside the Revolution at Etsy."

220 Etsy recently raised its fees: Matt Stieb, "Long vs. Short: Can Etsy Continue to Own the Handmade Space?," *New York*, http://

nymag.com/intelligencer/2018/11/can-etsy-continue-to-own-the-handmade-space.html.

221 rated Etsy 9 out of 10: Gallagher, "ESG Investing: Is Etsy a Responsible Investment?"

221 the stock price under his leadership: The price fell from $30.00 on April 16, 2015 to $11.39 on May 2, 2017, when Silverman was announced as CEO. It then rose to $54.39 as of February 20, 2020. Retrieved from Etsy, https://investors.etsy.com/stock-info/stock-info/default.aspx.

221 "There's only so much wiggle room": Gelles, "Inside the Revolution at Etsy."

221 In the words of *The Cut*: Larocca, "Etsy Wants to Crochet Its Cake, and Eat It Too."

222 "not in the shareholders' interest": Chad Dickerson, interview by Michael O'Leary, December 2, 2019. Unless otherwise noted, all further remarks from Dickerson are from this interview.

223 CEOs who focus: Shannon Roddel, "Heads-up, CEOs—Corporate Social Responsibility May Get You Fired, Study Finds," Phys.org, October 10, 2017, https://phys.org/news/2017-10-heads-up-ceoscorporate-social-responsibility.html.

223 "They could do it or they could not do it": Andrew Kassoy, interview by Michael O'Leary, February 5, 2020. Unless otherwise noted, all further remarks from Kassoy are from this interview.

223 "wish we were a B Corp": Ryan Honeyman and Tiffany Jana, *The B Corp Handbook: How You Can Use Business as a Force for Good* (Oakland: Berrett-Koehler, 2019), 54.

224 56 percent of Etsy's employees: Etsy, *Unlocking Opportunity: The Path to Owning "Special,"* Investor Day presentation, March 7, 2019, https://s22.q4cdn.com/941741262/files/doc_presentations/2019/03/Etsy-Investor-Day-Presentation_WITHFONTS_01.pdf.

225 the $428 billion: "Charitable Giving Statistics," National Philanthropic Trust, March 27, 2020, https://www.nptrust.org/philanthropic-resources/charitable-giving-statistics/.

225 the $4.4 trillion budget: "Budget," Congressional Budget Office, https://www.cbo.gov/topics/budget.

225 If they went beyond: Byrne, "False Profits," 44.

CHAPTER 9: CITIZEN CAPITALISM

227 After growing: Starbucks, "Starbucks Company Timeline" through 2015, https://www.starbucks.com/about-us/company-information/starbucks-company-timeline.

227 "opening its newest location": "New Starbucks Opens in Rest Room of Existing Starbucks," The Onion, June 27, 1998, https://www.theonion.com/new-starbucks-opens-in-rest-room-of-existing-starbucks-1819564800.

227 Starbucks opened eighty-four: "Starbucks Restructures Australian Operations to Position Business for Long-Term Growth," Starbucks Stories & News, accessed January 22, 2020, https://stories.starbucks.com/stories/2008/starbucks-restructures-australian-operations-to-position-business-for-long/; Abhishek Arora, "3 Reasons Starbucks Failed in Australia," Medium, January 15, 2019, https://medium.com/@arora.abhi/4-reasons-starbucks-failed-in-australia-f9efb125faeb; Ashley Turner, "Why There Are Almost No Starbucks in Australia," CNBC, July 25, 2018, https://www.cnbc.com/2018/07/20/starbucks-australia-coffee-failure.html.

229 "I set out": "Message from Howard Schultz to Partners: Onward with Love," Starbucks Stories & News, June 4, 2018, https://stories.starbucks.com/stories/2018/message-from-howard-schultz-to-partners-onward-with-love/.

229 "For Australians who want": David Marr, "I Feared the Worst, but Starbucks' New Flat White Is Actually Not Bad," *The Guardian*, January 6, 2015, https://www.theguardian.com/lifeandstyle/2015/jan/06/starbucks-flat-white-australia-coffee.

231 It's now selling it: Ashley Carman, "Burger King's Nationwide Rollout of the Impossible Whopper Starts Next Week," The Verge, August 1, 2019, https://www.theverge.com/2019/8/1/20750704/burger-king-nationwide-rollout-impossible-whopper-august-8.

231 McDonald's: Jordan Valinsky, "The Beyond Meat Burger Is Coming to McDonald's in a Canadian Test," CNN, September 26, 2019, https://www.cnn.com/2019/09/26/business/beyond-meat-mcdonalds-test/index.html.

231 White Castle, and Carl's Jr.: Carman, "Burger King's Nationwide Rollout."

231 10 billion land animals are killed for food: Doris Lin, "How Many Animals Do Human Kill Each Year?," ThoughtCo., January 7, 2019, updated October 23, 2019, accessed January 22, 2019, https://www.thoughtco.com/how-many-animals-are-killed-each-year-127631.

231 Together, these animals account: "How Does Eating Meat Harm the Environment?," PETA.org., n.d., accessed January 22, 2020, https://www.peta.org/about-peta/faq/how-does-eating-meat-harm-the-environment/; Lisa Friedman, Kendra Pierre-Louis, and Somini Sengupta, "The Meat Question, by the Numbers," *New York Times*, January 25, 2018, https://www.nytimes.com/2018/01/25/climate/cows-global-warming.html.

231 such as the Amazon: Dom Phillips, Daniel Camargos, Andre Campos, Andrew Wasley, and Alexandra Heal, "Revealed: Rampant Deforestation of Amazon Driven by Global Greed for Meat," *The Guardian*, July 2, 2019, https://www.theguardian.com/environment/2019/jul/02/revealed-amazon-deforestation-driven-global-greed-meat-brazil; Rhett Butler, "Cattle Ranching's Impact on the Rainforest," Mongabay.com., July 22, 2012, https://rainforests.mongabay.com/0812.htm.

231 Companies must compete: Assuming, of course, that government is ensuring fair competition.

231 The average profit margin: Mark J. Perry, "The General Public Thinks the Average Company Makes a 36% Profit Margin, Which Is About 5X Too High," American Enterprise Institute, https://www.aei.org/carpe-diem/the-public-thinks-the-average-company-makes-a-36-profit-margin-which-is-about-5x-too-high-part-ii/.

231 two in five Americans: Anna Bahney, "40% of Americans Can't Cover a $400 Emergency Expense," CNN Money, May 22, 2018, https://money.cnn.com/2018/05/22/pf/emergency-expenses-household-finances/index.html.

232 our values in surveys: Timothy M. Devinney, Pat Auger, and Giana M. Eckhardt, *The Myth of the Ethical Consumer* (Cambridge: Cambridge University Press, 2010).

232 "Almost all of our customers": Andrew Edgecliffe-Johnson,

"Beyond the Bottom Line: Should Business Put Purpose Before Profit?," *Financial Times*, January 4, 2019.

232 Yet 70 percent: Kimberly Amadeo, "Components of GDP Explained," The Balance, December 16, 2019, https://www.thebalance.com/components-of-gdp-explanation-formula-and-chart-3306015.

232 The average household spends: US Bureau of Labor Statistics, "Consumer Expenditures—2018," September 10, 2019, https://www.bls.gov/news.release/pdf/cesan.pdf.

232 LEED-certified buildings: U.S. Green Building Council, "Benefits of Green Building," press release, May 24, 2019, https://new.usgbc.org/press/benefits-of-green-building.

232 In commercial real estate: Thomas P. Lyon et al., "CSR Needs CPR: Corporate Sustainability and Politics," *California Management Review* 60, no. 4 (August 2018), https://www.hbs.edu/faculty/Publication%20Files/Lyon_et_al_2018_CMR_f4406d48-0511-4f2a-a83f-7ee1e2952c8c.pdf.

232 By 2040, more than half: Peter Valdes-Dapena, "By 2040, More Than Half of New Cars Will Be Electric," CNN, September 6, 2019, https://edition.cnn.com/2019/05/15/business/electric-car-outlook-bloomberg/index.html.

232 they may become less expensive: Minda Zetlin, "Here's Why the Next Car You Buy Will Be Electric," *Inc.*, April 20, 2019, https://www.inc.com/minda-zetlin/electric-car-ev-battery-range-cost-less-than-gas.html.

232 Drivers keep a new car: "Buying a Car: How Long Can You Expect a Car to Last?," Autotrader, July 15, 2019, https://www.autotrader.com/car-shopping/buying-car-how-long-can-you-expect-car-last-240725.

233 Walmart is now the largest: Jennifer Chait, "Largest Organic Retailers in North America," The Balance, November 20, 2019, https://www.thebalancesmb.com/organic-retailers-in-north-america-2011-2538129.

234 "throwaway living": Yesh Pavlik Slenk, "Can the Circular Economy End the Era of 'Throwaway Living'?," Environmental Defense Fund, June 20, 2019, http://business.edf.org/blog/2019/06/20/can-the-circular-economy-end-the-era-of-throwaway-living?.

234 the average American home: Margot Adler, "Behind the Ever-Expanding American Dream House," NPR, July 4, 2006, https://www.npr.org/templates/story/story.php?storyId=5525283.

234 "slow clothing" movement: This is an attempt to reduce the fast-fashion footprint.

234 Sustainable fashion is now: Maggie Kohn, "Retailers Are Betting on Sustainable Fashion to Drive Sales," TriplePundit, June 24, 2019, https://www.triplepundit.com/story/2019/retailers-are-betting-sustainable-fashion-drive-sales/84016.

234 they staged a walkout: Tina Casey, "The Wayfair Effect: Companies Find Their Voices Rebuking Racism," TriplePundit, July 31, 2019, https://www.triplepundit.com/story/2019/wayfair-effect-companies-find-their-voices-rebuking-racism/84416.

234 Wayfair had signed: Kate Trafecante and Nathaniel Meyersohn, "Wayfair Workers Plan Walkout in Protest of Company's Bed Sales to Migrant Camps," CNN, June 26, 2019, https://www.cnn.com/2019/06/25/business/wayfair-walkout-detention-camps-trnd/index.html.

234 "we believe it is": Ibid.

235 "no part in enabling": Ibid.

235 "really the appropriate channel": Sarah Spellings, "What Happens After the Wayfair Walkout," *The Cut*, June 27, 2019, https://www.thecut.com/2019/06/what-happened-at-the-wayfair-employee-walkout.html.

235 "We cannot outsource": Quoted in Scott Shane and Daisuke Wakabayashi, "'The Business of War': Google Employees Protest Work for the Pentagon," *New York Times*, April 4, 2018, https://static01.nyt.com/files/2018/technology/googleletter.pdf.

235 Sixty-seven percent of employees: George Serafeim, "Investors as Stewards of the Commons?," *Journal of Applied Corporate Finance* 30, no. 2 (Spring 2018): 20.

236 nearly half of millennials: Andy Hames, "The Rise of Employee Activism: Lessons from the Wayfair Walkout," *Employee Benefit News*, July 2, 2019, https://www.benefitnews.com/opinion/wayfair-walkout-exemplifies-importance-of-company-culture.

236 65 percent of Gen Xers: Bank of America Corporation, "2018

U.S. Trust Insights on Wealth and Worth," June 2018, https://newsroom.bankofamerica.com/system/files/2018_US_Trust_Insights_on_Wealth_and_Worth_Overview.pdf.

237 more than $12 trillion: Edgecliffe-Johnson, "Beyond the Bottom Line."

237 This has grown fourfold: Ibid.

237 As You Sow: "Invest in Your Values," As You Sow, November 10, 2018, https://www.asyousow.org/invest-your-values.

237 Triodos Bank: Triodos Bank website, https://www.triodos.com/.

239 In the 2016 presidential election: Drew DeSilver, "U.S. Trails Most Developed Countries in Voter Turnout," Fact Tank, Pew Research Center, https://www.pewresearch.org/fact-tank/2018/05/21/u-s-voter-turnout-trails-most-developed-countries/.

239 Others are focused on: Karl Evers-Hillstrom, "Lobbying Spending Reaches $3.4 Billion in 2018, Highest in 8 Years," OpenSecrets.org, January 25, 2019, https://www.opensecrets.org/news/2019/01/lobbying-spending-reaches-3-4-billion-in-18/.

239 30 percent of those: Gary Gutting, "Is Voting Out of Self-Interest Wrong?," *New York Times*, March 31, 2016, https://opinionator.blogs.nytimes.com/2016/03/31/is-voting-out-of-self-interest-wrong/.

241 "A social organism": William James, *The Will to Believe: And Other Essays in Popular Philosophy* (New York, London, and Bombay: Longmans Green and Co., 1907), 24.

241 Three in four people: Miguel Padró, "Unrealized Potential: Misconceptions About Corporate Purpose and New Opportunities for Business Education," The Aspen Institute, May 28, 2014, https://assets.aspeninstitute.org/content/uploads/files/content/docs/pubs/Aspen%20BSP%20Unrealized%20Potential%20May2014v.2.pdf?_ga=2.40107429.603537148.1576800305-420537432.1576800305, 2.

241 the proportion of American teenagers: Paul Collier, *The Future of Capitalism: Facing the New Anxieties* (New York: HarperCollins, 2018), 45.

241 "Trust is really": Halla Tómasdóttir, interview by Michael O'Leary, October 31, 2019. Unless otherwise noted, all further remarks from Tómasdóttir are from this interview.

242 Tómasdóttir's organization is called The B Team: If you're finding the nomenclature around B Corps, B Lab, B Team, and Plan B all a bit confusing, you have officially joined the world of responsible capitalism.

242 "Today, amidst rising": Sharan Burrow, Paul Polman, and Halla Tómasdóttir, "Realizing Our Ambition Requires Courage and Accountability," The B Team, October 15, 2019, https://bteam.org/our-thinking/thought-leadership/realizing-our-ambition-requires-courage-and-accountability.

242 includes many giants: David Crane, the former head of NRG whom we profiled in chapter 5, is also a member.

242 Chobani not only pays: Simon Mainwaring, "Purpose at Work: How Chobani Builds a Purposeful Culture around Social Impact," *Forbes*, August 27, 2018, https://www.forbes.com/sites/simonmainwaring/2018/08/27/how-chobani-builds-a-purposeful-culture-around-social-impact/.

242 Danone has given: Emmanuel Faber, "One Person, One Voice, One Share: Time to Let Our People Go Create and Own the Future," LinkedIn, April 26, 2018, https://www.linkedin.com/pulse/one-person-voice-share-time-let-our-people-go-create-own-faber.

EPILOGUE: A CAPITALIST REFORMATION

245 "If we are not": Steven Mufson, "Jim Rogers, Duke Energy Executive Who Promoted Clean Energy, Dies at 71," *Washington Post*, December 21, 2018, https://www.washingtonpost.com/local/obituaries/jim-rogers-duke-energy-executive-who-promoted-clean-energy-dies-at-71/2018/12/21/53b0f1f4-04ad-11e9-b5df-5d3874f1ac36_story.html.

245 "using his perch": Brian Murray, "Remembering Duke Energy's Jim Rogers: A 'Cathedral' Thinker," *Forbes*, January 8, 2019, https://www.forbes.com/sites/brianmurray1/2019/01/08/remembering-a-cathedral-thinker/#487dcdf5196d.

245 "led by example": Fred Krupp (@FredKrupp), "Sorry to hear the news of Jim Rogers' passing," Twitter, December 18, 2018, 3:22 p.m., https://twitter.com/FredKrupp/status/1075169563262750720.

245 He was the first prominent: Mufson, "Jim Rogers."

246 "this little pissant utility CEO": Tom Williams, interview by Michael O'Leary, October 21, 2019.

246 In 2001, he shocked: Julie Creswell, "James Rogers, 71, Dies; Utility Chief and Clean Energy Advocate," *New York Times*, December 21, 2018, https://www.nytimes.com/2018/12/21/obituaries/james-rogers-dead.html.

246 Between 2005 and 2018, Duke Energy: Duke Energy, "Jim Rogers, former Duke Energy chairman and CEO, dies at age 71," December 18, 2018, https://news.duke-energy.com/releases/jim-rogers-former-duke-energy-chairman-and-ceo-dies-at-age-71.

246 Some saw Rogers: "44 Arrested for Protesting Duke's Climate Hypocrisy," Greenpeace, July 6, 2010.

246 "a genuine anomaly": Clive Thompson, "A Green Coal Baron?," *New York Times Magazine*, June 22, 2008, https://www.nytimes.com/2008/06/22/magazine/22Rogers-t.html.

246 Duke cut the proportion: Creswell, "James Rogers, 71, Dies."

246 "Rogers's environmentalism is": Thompson, "A Green Coal Baron?"

247 "a long-term perspective": Murray, "Remembering Duke Energy's Jim Rogers."

247 "balance all of that": Tom Williams, interview by Michael O'Leary, October 21, 2019.

248 "Few trends": Milton Friedman, *Capitalism and Freedom*, 40th Anniversary Edition (Chicago: University of Chicago Press, 2002 [1962]), 133.

248 more than 60 percent: Samuel Bowles, *The Moral Economy: Why Good Incentives Are No Substitute for Good Citizens* (New Haven: Yale University Press, 2016), 70.

248 *only a third*: Ibid., 44.

248 We donated: "*Giving USA 2019*: Americans Gave $427.71 Billion to Charity in 2018 Amid Complex Year for Charitable Giving," Giving USA Foundation, June 18, 2019, https://givingusa.org/giving-usa-2019-americans-gave-427-71-billion-to-charity-in-2018-amid-complex-year-for-charitable-giving/.

248 on a dollar basis: "Table17-0336—Effective Marginal Tax Benefit of Charitable Contributions Under Current Law and Conference Agreement for H.R. 1, the Tax Cuts and Jobs Act; by Expanded

Cash Income Percentile, 2018," Tax Policy Center, December 17, 2017, https://www.taxpolicycenter.org/model-estimates/charitable-contributions-and-tcja-nov-2017/t17-0336-effective-marginal-tax-benefit.

248 nearly triple the national percentage: Karl Zinsmeister, "Statistics," Philanthropy Roundtable *Almanac*, https://www.philanthropyroundtable.org/almanac/article/statistics.

249 "By propagating ideologically inspired": Sumantra Ghoshal, "Bad Management Theories Are Destroying Good Management Practices," *Academy of Management Learning & Education* 4, no. 1 (March 2005): 76.

249 "we label as 'irrational'": Luigi Zingales, "Does Finance Benefit Society?," presidential address, American Finance Association Annual Meeting, Boston, MA, January 4, 2015, https://faculty.chicagobooth.edu/luigi.zingales/papers/research/finance.pdf, 32.

249 granted in business each year: US Department of Education, National Center for Education Statistics, "Fast Facts," *Digest of Education Statistics, 2017* (Washington, DC: NCES 2018-070, 2019), chapter 3, https://nces.ed.gov/fastfacts/display.asp?id=37.

249 "Our theories and ideas": Ghoshal, "Bad Management Theories," 75.

249 Teaching economics makes: Zingales, "Does Finance Benefit Society?," 32; Jeffrey Pfeffer, "Why Do Bad Management Theories Persist? A Comment on Ghoshal," *Academy of Management Learning & Education* 4, no. 1 (March 2005): 97.

250 "When everything that matters": John C. Bogle, *The Battle for the Soul of Capitalism* (New Haven: Yale University Press, 2005), 4.

250 "low-wage clerk": David Foster Wallace, "This Is Water," commencement speech, Kenyon College, Gambier, OH, 2005, Farnam Street, https://fs.blog/2012/04/david-foster-wallace-this-is-water/.

251 "How selfish soever": Vernon L. Smith and Bart J. Wilson, eds., *Humanomics: Moral Sentiments and the Wealth of Nations for the Twenty-First Century* (Cambridge: Cambridge University Press, 2019), 8–9, quoting the first sentence of Adam Smith's *The Theory of Moral Sentiments* (1759).

251 "The wise and virtuous man": Quoted in Dominic Barton, "Capi-

talism for the Long Term," *Harvard Business Review*, March 2011, 7; *The Theory of Moral Sentiments* is available online at http://knarf.english.upenn.edu/Smith/tmstp.html.

251 "How can it be": Pope Francis, "Evangelii Gaudium" (November 24, 2013), quoted in Laurie Goodstein and Elisabetta Povoledo, "Pope Sets Down Goals for an Inclusive Church, Reaching Out 'on the Streets,'" *New York Times*, November 26, 2013, https://www.nytimes.com/2013/11/27/world/europe/in-major-document-pope-francis-present-his-vision.html.

253 "If you go to Europe": *"Short-Termism in Financial Markets," Hearing Before the Subcommittee on Economic Policy of the Committee on Banking, Housing, and Urban Affairs, US Senate,* 111th Cong., 2d Sess., April 29, 2010, https://www.govinfo.gov/content/pkg/CHRG-111shrg61654/pdf/CHRG-111shrg61654.pdf.

253 "Their vision was as clear": Jim Rogers, "Build Your Own Cathedral," speech delivered at Queen's University, Charlotte, NC, May 4, 2007, emailed to authors by Duke Energy.

253 In 1850, the tallest building: Trinity Church Wall Street, "History," December 22, 2013, https://www.trinitywallstreet.org/about/history.

253 none of the tallest buildings: John Emslie, *Comparative View of the Principal Buildings in the World*, engraving (London: James Reynolds, March 30, 1850), https://i.redd.it/f2cbvt5a80j01.jpg.

253 "The man who builds": Library of Congress, "Prosperity and Thrift: The Coolidge Era and the Consumer Economy, 1921–1929," February 1, 2002, https://www.loc.gov/teachers/classroommaterials/connections/prosperity-thrift/thinking3.html.

254 "Make no little plans": Patrick T. Reardon, "Burnham Quote: Well, It May Be," *Chicago Tribune*, January 1, 1992, https://www.chicagotribune.com/news/ct-xpm-1992-01-01-9201010041-story.html.

254 "We must take the same approach": "Duke CEO Calls for 'Cathedral Thinking' to Become Energy Efficient," *Natural Gas Intelligence*, November 5, 2007, https://www.naturalgasintel.com/articles/17327-duke-ceo-calls-for-cathedral-thinking-to-become-energy-efficient.

INDEX

Key to abbreviations: CSR = corporate social responsibility; ESG = environmental, social, and governance programs

ABOUT THE AUTHORS

MICHAEL O'LEARY was on the founding team of Bain Capital's social impact fund. Previously, he invested in consumer, industrial, and technology companies through Bain Capital's private equity fund. He has served as an economic policy adviser in the United States Senate and on two presidential campaigns. Michael studied philosophy at Harvard College and earned his MBA from the Stanford Graduate School of Business. He lives in New York.

WARREN VALDMANIS leads a social impact fund that invests in the American workforce. He was previously a managing director with Bain Capital's social impact fund, and before that invested with Bain Capital's private equity team for over a decade. He grew up in Canada and has lived and worked in Australia, Chile, France, Hong Kong, Japan, South Africa, and the United States. Warren studied economics at Dartmouth College and earned his MBA from Harvard Business School. He lives with his wife, Kristin, and four children in Portland, Maine.